GW01182706

In the fiscal crises that faced the governments of many countries in the developing world during the 1980s there were those who argued for a greater reliance on the implementation and collection of user fees for financing irrigation operation and maintenance. In this book, Leslie Small and Ian Carruthers examine in detail the potentials and limitations of user fees, combining their extensive field experience in irrigation in developing countries with simple concepts of economics to propose possible institutional and financial reforms which would not simply ask farmers to pay for an inadequate irrigation service, but would create the potential for significant improvements in the quality of the service provided. The proposed elements of any such reform are examined in detail – a system of user fees covering the recurrent costs of irrigation; a financially autonomous irrigation agency that can retain and use the fees to operate and maintain the irrigation facilities; and a macro policy environment that is not unduly skewed against the agricultural sector.

Farmer-financed irrigation

Farmer-financed irrigation: the economics of reform

LESLIE E. SMALL

Associate Professor, Department of Agricultural Economics
Cook College, Rutgers University

IAN CARRUTHERS

Professor of Agrarian Development
Wye College, University of London

Published in association with the
International Irrigation Management Institute

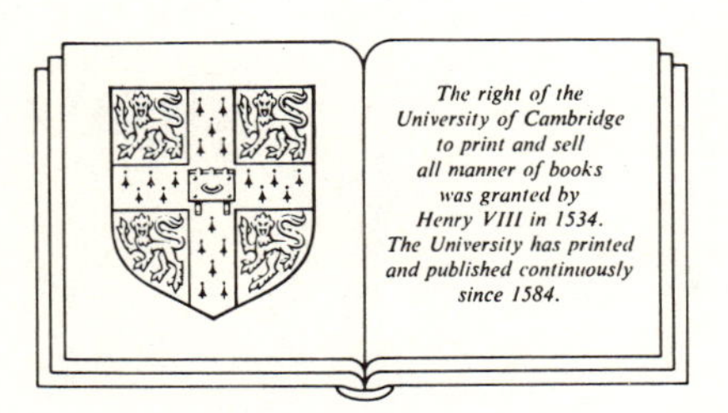

CAMBRIDGE UNIVERSITY PRESS

Cambridge

New York   Port Chester   Melbourne   Sydney

Published by the Press Syndicate of the University of Cambridge
The Pitt Building, Trumpington Street, Cambridge CB2 1RP
40 West 20th Street, New York, NY 10011-4211, USA
10 Stamford Road, Oakleigh, Melbourne 3166, Australia

First published 1991

Printed in Great Britain at the University Press, Cambridge

British Library cataloguing in publication data

Small, Leslie E.
Farmer-financed irrigation: the economics of reform.
1. Agricultural land. Irrigation. Economics
I. Title II. Carruthers, I. D. (Ian Douglas)
338.162

Library of Congress cataloguing in publication data

Small, Leslie E.
Farmer-financed irrigation: the economics of reform / Leslie E. Small and Ian Carruthers.
p. cm.
Includes index.
ISBN 0 521 38073 1
1. Irrigation – Developing countries – Finance. I. Carruthers, Ian D.
II. Title.
HD1741.D44S59 1991
333.91′3′091724 – dc20 91-8046 CIP

ISBN 0 521 38073 1 hardback

PN

Contents

Preface

For many years we have been interested in and have worked on problems of irrigation management in Third World countries. During the mid 1980s, both of us came to focus our attention on the increasingly difficult problems that governments faced with respect to financing the costs (especially the recurrent costs) of irrigation. Small, during a two-year leave from Rutgers University spent at the newly established International Irrigation Management Institute (IIMI) in Sri Lanka, undertook an extensive study of irrigation financing policies that included a literature review and case studies of Indonesia, South Korea, Nepal, the Philippines and Thailand. The study was funded by the Asian Development Bank as part of a Regional Technical Assistance activity. At about the same time Carruthers was undertaking a similar study, funded by the United States Agency for International Development (USAID) and managed by Devres Inc. of Washington D.C., which included case studies of policies in the Dominican Republic, Indonesia, Morocco, Peru and the Philippines. Both of these studies reflected the concern of international lending and donor agencies, as well as national agencies, that financial difficulties were becoming a serious constraint on irrigation performance.

Our studies have convinced us that financial policies can have important effects on irrigation performance. We are also convinced that economic perspectives and insights provide a powerful approach to the analysis of alternative financial policies of irrigation – an approach which we believe can be appreciated and used by policy makers and irrigation managers regardless of the nature of their specialised training. It is these considerations that have led us to write this book.

Throughout the book we draw liberally from our original studies; however, to avoid tedious repetition of references, we generally do not cite them in the body of the text. The findings of the studies that Small worked on were originally presented in:

> Leslie E. Small, Marietta S. Adriano and Edward D. Martin (1986). 'Regional Study on Irrigation Service Fees: Final Report' (2 vol). Submitted to the Asian Development Bank by the International Irrigation Management Institute, Digana, Sri Lanka.

More recently, they have been published together with some related case studies in:

> Leslie E. Small, Marietta S. Adriano, Edward D. Martin, Ramesh Bhatia, Young Kun Shim and Prachandra Pradhan (1989). *Financing Irrigation Services: A Literature Review and Selected Case Studies from Asia*. Colombo, Sri Lanka: International Irrigation Management Institute.

The reference to those that Carruthers worked on is:

> Devres, Inc. 'Irrigation Pricing and Management'. Report submitted to the U.S. Agency for International Development by Ian Carruthers *et al.*, Contract No. OTR-0091-C-00-4466-00.

Financial support for this study has come from a number of institutions. Funding support for the original case studies came from the Asian Development Bank and IIMI, and from the United States Agency for International Development (USAID). Our home institutions, Cook College and the New Jersey Agricultural Experiment Station of Rutgers University, and Wye College of the University of London, have provided their support during the writing phase. Our collaboration on this book required us to spend time together at key points in the preparation of the manuscript. We are grateful to the Irrigation Support Project for Asia and the Near East (ISPAN), sponsored by the United States Agency for International Development (USAID), for providing funds to use for the necessary travel.

Many individuals have provided assistance, ideas and information which, in one way or another, have helped to make this book possible. We are grateful to them for their many forms of assistance. We particularly want to acknowledge our co-workers and colleagues in the original case studies: Marietta Adriano, Edward Martin, Effendi Pasandaran, Prachandra Pradhan and Young Kun Shim in the IIMI Group and N. S. Peabody III, A. A. Bishop, A. D. Le Baron, Rekha Mehra, Ramchand Oad and Dennis A. Wood in the Devres Group. The late Dean Peterson of Utah State University was also a full participant and an extremely wise counsellor to this team. We are grateful to the many people, including government officials and farmers, who provided us with information and who took the time to answer our many questions during the individual case studies.

During the spring semester of 1990, Small had the opportunity, while on sabbatical leave from Rutgers University, to use a preliminary draft of the manuscript in teaching an interdisciplinary graduate course at Cornell University entitled 'Socio-Technical Aspects of Irrigation'. The lively and thoughtful discussions that ensued helped shape the revision of the manuscript. We wish to thank all the faculty and students who participated in the course, with a special word of thanks (without implicating them in the final product) to Professors Randy Barker, Norman Uphoff and Mike Walter. At Cornell we also wish to thank the Department of Agricultural Economics and the International Agriculture Program for logistic and financial assistance that made the teaching arrangement possible. Finally, we should acknowledge financial assistance from IIMI who have followed their mandate to disseminate research findings. This has enabled wider distribution of the text in the developing world and also its use in the Wye External Programme distance learning course entitled Water Resource Economics.

For the helpful word processing assistance during the many drafts of this manuscript, we wish to thank Estelle Scaiano and Mary Arnold.

Both the field work and the writing of the manuscript have taken us away from our families all too much over the past several years. We owe a large debt of thanks to them for their forbearance and support.

Leslie Small
Ian Carruthers

1

Irrigation financing in perspective

1.1 *Irrigation in the context of Third World development*

1.1.1 *Irrigation and world food supplies*

Irrigation provides supplementary water supply to one-fifth of the world's cultivated land, from which one-third of the world's food is harvested. Many of the world's poorest people are dependent on this food. Billions of low-income people struggle to supplement inadequate and unreliable rainfall with irrigation.

The stakes are clearly high. Two statistics highlight this. One in five of all people in the world is a Chinese peasant and most of them are irrigation farmers. Every month there are a million more Indian farmers and most are or would like to become irrigation farmers.

Irrigation is a potentially effective investment to service the basic needs for food and employment in the developing world. But the investment necessary to develop new irrigation systems is costly. And the expense does not end with the construction of irrigation facilities. The provision of reliable irrigation service requires recurrent expenditures for operation and maintenance.

Irrigation has been an extremely important development investment area in recent years and it is going to be even more important in the future. In several large developing countries like China, India, Indonesia and Pakistan, half of all agricultural investment goes into irrigation. Some 25–30% of World Bank agricultural lending is allocated to irrigation. In the next 10 years between $50 and $120 billion will be spent on new irrigation and on rehabilitating existing projects.

These investments reflect a dramatic increase in the potential returns to irrigation brought about by important technological changes in agricultural production. These changes (collectively known as the 'green revolution') were centred upon the widespread adoption of new stiff-strawed varieties of wheat and rice that responded to high doses of artificial nitrogenous fertiliser. The high potential yield of this seed-fertiliser technology was only obtainable with crop protection, including an adequate, reliable supply of soil moisture. This technology spread first to areas with good irrigation, and provided an impetus for further irrigation development. As a result, nearly three-quarters of recent increases in agricultural production have come from irrigated land.

The green revolution was thus centred on a package of modern scientific inputs that has pushed grain production further from the traditional subsistence methods into the cash economy. On the horizon in the near future is a new set of seeds and plants that will be the product of biotechnology. We can be fairly confident that although new technology will increase potential returns, the total variable costs will also rise over time. All food crops will in time become cash crops to a greater or lesser degree.

Irrigation will continue to be important in providing the secure growing conditions that will make high input, high output farming economically feasible. Any failure of irrigation to function in line with its potential implies extremely high opportunity costs in economic and human terms, as the scope for rural poverty alleviation would be very much reduced.

1.1.2 *Irrigation problems*

Unfortunately the consensus among irrigation researchers and financing agencies is that irrigation is not performing anywhere near its potential. As one reviewer concluded in a damning summary of field evidence:

> Evaluations of public irrigation systems have shown that, in most, service has deteriorated due to faulty design and construction, neglected maintenance, and inefficient operation. Distribution channels, if aligned properly to begin with, become silted up or breached as time goes by. Even in systems designed for regular rotational water distribution, deliveries to most farmers are erratic and unreliable.[1]

There is considerable evidence that the potential gains from irrigation are far from being fully realised. For example, inadequate water management is held to be the largest single factor in explaining the gap between actual and potential rice yields. It is estimated that more than half the water supply lost before reaching the crops could, with sound infrastructure and good management, be beneficially used, increasing water availability for crop growth by up to 25%.

Numerous interrelated reasons account for the failure of irrigation investments to produce their intended benefits. No single reason, not even financial problems, can be put forward to explain failure of irrigation investments to realise their maximum potential. Problems cited in various analyses include:

(i) inadequate preparation of projects (e.g. poor assessment of water availability, soil analysis, etc.) and faulty design (especially at the farm end of systems);
(ii) substandard, careless construction;
(iii) underinvestment in infrastructure (e.g. lack of drainage, insufficient control structures);
(iv) poor canal management and organisation (e.g. faulty personnel policies);
(v) insufficient financial resources and priority for operation and maintenance;
(vi) poor crop production techniques and agricultural services (e.g. use of low quality seeds, no or inadequate extension services);
(vii) neglect of public health aspects of irrigation design and operation;
(viii) poor land levelling and water management at the farm level;
(ix) exogenous problems such as unrealistically low prices resulting from crop pricing policy and unreliable delivery of inputs such as fertiliser or electricity;
(x) poor coordination between engineers and agricultural specialists (Box 1.1).

These problems are interlinked. One problem can initiate another which can cause a third and so forth. Poor canal design can lead to shortage of water. In turn, this leads to farmers adopting unorthodox coping mechanisms or even stealing extra supplies which, in arid areas, will cause waterlogging at the head of canals and drought and soil salinity in the irrigated lands at the tails. Low returns to farmers in these circumstances may, in time, lead to farmer refusal to pay irrigation charges or service fees. Financial delinquency by a few farmers may rapidly lead to widespread non-payment and starve the operating agency of financial resources which may in turn affect operation and maintenance.

The focus of this book is on the financial problems of irrigation. But as the above paragraph demonstrates, irrigation is part of an interdependent socioeconomic system, and therefore reform of the financing com-

---

BOX 1.1

Phases of interdisciplinary cooperation in the history of developing-country irrigation

Four phases of interdisciplinary cooperation can be identified according to the specialties involved:

Phase I **Military engineers, civil engineers, administrative officers, and financial analysts**

Phase II **Civil engineers, administrators, financial analysts and *agriculturalists***

Phase III **Civil engineers, financial analysts, agriculturalists and *economists***

Phase IV **Civil engineers, financial analysts, agriculturalists, economists and *farmers* (plus specialists in other areas such as public health, the environment, and sociology).**

Phase I lasted the longest and Phase II did not really dawn until the second half of the twentieth century. Phase III is a post 1950s phenomenon and Phase IV is yet to appear. The neglect of agriculture and agriculturalists is longstanding. Consider the following citation from the *Indian Agriculturalist* of July 1876.

> **There is great truth in his (Corbett's) assertion that an irrigation cry and a drainage cry, have induced the Government to embark in projects purely engineering and not agricultural, to trust the agricultural education of India solely to engineers and to district officers; the former of whom look upon agricultural projects from a purely engineering point of view, while the latter have little interest in agricultural matters beyond the narrow one of collecting the revenue. In a country which is so largely dependent as India not only for the subsistence of its vast population but for its political maintenance, upon the productiveness of the earth, the science of agriculture should doubtless be made of the first importance and should have been called in to aid all projects of agricultural improvement.[2]**

---

ponent requires a holistic approach that recognises the complexity of interrelationships among all the components of the system.

1.1.3 *Macroeconomic setting*

The severity of the present economic and financial crisis facing most developing countries and the prolonged international recession of the 1980s are generating unprecedented difficulties for governments in general, and, in our context, for irrigation authorities in particular. The current macroeconomic context has inevitably created new problems and priorities for the irrigation sector. Typically, national debt service obligations are causing extreme obstacles: many public sector institutions have liquidity crises and some agencies are practically insolvent. A precondition for any effective irrigation sector policy analysis for the 1990s is a consideration and proper understanding of the macroeconomic framework. This is particularly true of any analysis of financial elements of irrigation policy. Unless and until the ramifications of the macro-economy are recognised, there can be no guarantee that any change in irrigation policy will be an effective, let alone an efficient, improvement. In an unstable economy, a policy change that would be helpful in other circumstances could even cause damage.

Despite substantial economic progress in developing countries during the 1970s, the varied external and internal economic shocks of the 1980s have revealed crucial structural weaknesses in these economies. It is now apparent that growth in the 1970s was being obtained at high investment cost. In many areas such as manufacturing (and in some sections of agriculture), high levels of protection and public sector subsidy using inappropriate trade, industrial, financial and exchange rate policies led to sheltered investments in activities where many developing countries lacked a clear comparative advantage.

In the agricultural sector, government market regulation and input and output price controls, together with archaic institutional frameworks, limited the capacity of the sector to benefit fully from the general economic growth of the 1970s. There is now belated but widespread recognition of the negative impacts that taxing agriculture to fund urban sector needs can have on a nation's economic growth; however, the temptation will remain in some countries to continue such policies because of the scale of agriculture, the severity of the adjustment problem, and the few alternative policy instruments available to governments.

If agriculture did well in some developing countries in the 1970s (such as where irrigated wheat and, latterly, rice were the main crops) it often did so in spite of, rather than because of, public sector policy. It seems likely that agricultural planners in the 1990s will have to rely upon agricultural growth yet again to stimulate their economies but without moving the internal terms of trade too much in favour of agriculture.

By the mid-1980s many irrigation agencies and projects were facing unrelenting financial problems. We need to explore how this could have arisen when irrigation at least appeared to be a relatively successful technology during the 1970s. It is certain that the unfavourable macroeconomic situation played a key part.

The macroeconomic picture that emerged in the early 1980s was confused. However, recognition of government use of massive external borrowing and imports to bolster the gains of the 1970s and to sustain investment programmes in the face of unfavourable world economic conditions, sharpens the image. It is clear that despite public sector initiatives, economic growth failed to resume previous levels; high interest rates and global inflation prevailed; and for many countries there was a stagnation in terms of trade. The growth of external deficits was exacerbated in most developing countries by a fall in government revenues.

Many governments underestimated the severity and duration of recession and borrowed heavily for both investment and consumption purposes. The perception that structural adjustment to a new economic order was a necessity was only slowly realised in 1979–83. Furthermore, achieving stabilisation and adjustment, once the problem was recognised, has proved to be a harsh and costly process. In addition to unfavourable external factors, there are often domestic political imperatives such as the need to reduce the impact of urban unemployment and to protect infant industries not yet able to 'grow up' to competitive independence. Responding to these features will inevitably restrict or slow the adjustment process.

In this process, irrigation institutions have probably suffered less than manufacturing industry; however, the expansion of the area irrigated has generally slowed, and the farmers dependent upon technology such as pumped schemes and groundwater have often faced rapidly increased costs (or have added to the government's financial burden). The expenditure patterns of many governments between 1979 and 1985 have produced a medium-term shortage of financial resources. There are clear limits to the ability and willingness of many governments to finance irrigation infrastructure from general revenue. In our field studies we were repeatedly informed by government officials that financial stringency in public expenditure threatens to reduce further the generally unsatisfactory standards of irrigation performance.

Thus, the irrigation sector illustrates the general public sector recurrent cost problem: expansion of the investment portfolio resulting in large increases in the demand for recurrent expenditures to operate and maintain the infrastructure; and an inability to finance these expenditures adequately. The scope for continuation of many of the direct and indirect financial subsidies of the past is extremely limited. But to allow irrigation facilities to deteriorate at a time when complementary inputs have combined to create unprecedented productivity for irrigation would be irrational. Hence, most governments in developing countries are being forced to reconsider their policies toward farmer payments for and participation in irrigation operation and maintenance. Financing irrigation with funds provided by farmers through one means or another becomes nearly inevitable.

But while macroeconomic conditions have created fiscal stringencies that make governments look to increased funding of irrigation costs by the farmers, other broad economic forces may make this approach difficult. For example, the success of national and international efforts to increase agricultural production may have been great enough to depress crop prices.

This is well illustrated in the case of Indonesia. Between 1976 and 1983, Indonesia's rice and wheat imports averaged 2.6 million tonnes and cost about $500 million annually. The government has given high priority to intensive efforts to increase agricultural production. These efforts have included promotion of modern rice varieties with high levels of fertiliser application, and massive investments in rehabilitating and extending irrigation. Since 1968 the World Bank alone has provided more than one billion dollars for irrigation expansion and improvement. These policies have combined to produce a rice surplus at the favourable price environment presently enjoyed by farmers. However, the government is struggling to maintain high real producer prices because prospects for exports are very limited, the financial cost of crop purchase for government storage is extremely high, and the physical limits to suitable grain stores are nearly reached. If such circumstances combine to reduce farm prices and farm incomes, the scope for simultaneous significant increases in fees or charges for irrigation are much reduced. This illustrates the broad and complex context within which irrigation financing and water pricing policies have to be considered.

1.2 *The approach of the book*

1.2.1 *Focus on financial policies for irrigation*

Our work on irrigation problems in Third World countries over the past several years has convinced us of the importance of irrigation financing policies. Severe financial difficulties in the irrigation sector are common, often leading to declining irrigation performance.

These financial difficulties are related to the fact that public irrigation is often heavily subsidised. While such subsidies are commonly found in industrialised nations as well as in Third World countries, the financial difficulties are often greater in the latter nations because of greater overall budgetary constraints. As irrigation development in the Third World has proceeded over the past several decades, levels of subsidies that were acceptable when the total amount of irrigation was small have become increasingly burdensome to government budgets.

From a straightforward accounting view, the financial subsidies given to irrigation users are easy to identify. For Third World countries these subsidies almost always include not only the investment cost of the irrigation facilities, but also part or even all of the expenditure needed to pay operation and maintenance costs.

The existence of a financial subsidy, however, does not necessarily mean that a true economic subsidy is being given to the irrigation farmers, because of the myriad of indirect charges and implicit taxes that are levied on them. Governments in developing countries often squeeze irrigation (and other) farmers by manipulation of agricultural markets, export duties, and the maintenance of overvalued exchange rates. These typically add up to a massive financial burden to agricultural producers and exporters and subsidies to the mainly industrial importers. Furthermore, the squeezing of resources from agriculture by indirect means, negatively affecting the relative prices between the agricultural and industrial sectors (what economists call the domestic terms of trade between agriculture and industry), can have extremely harmful disincentive effects.

Public policy affects the availability and price of virtually all inputs and outputs in the irrigation sector. In evaluating a proposed policy change, such as, for instance, an increase in water fees to signal the real costs of providing irrigation service to farmers, the overall context has to be simultaneously considered; otherwise, infeasible or inappropriate policies may be advocated. This is perhaps best illustrated by a hypothetical but fairly typical example. If rice is the major crop in an irrigation system and the price is held at two-thirds of the free market or equilibrium price, then there is a transfer of income from producers (rural) to consumers (mainly urban). The rural areas will in effect be subsidising urban wage earners and urban industrialists, but it is difficult to determine by how much. To impose high irrigation charges in such circumstances in order to generate public savings, or even to cover the costs of installation or just operation and maintenance, may be impractical as well as unjust. However, if agriculturalists at the same time are subject to a set of subsidies and taxes for credit, fertiliser and other inputs, export taxes, export quotas and so forth, the policy environment becomes exponentially more complex. Irrigation authorities in many countries operate within just such a policy framework involving complex economic distortions.

Mobilising financial resources for irrigation is thus but one aspect of irrigation policy. Farmers are simply one possible source of finance. Finance is not the only resource farmers can offer: their labour may be of greater value. However, making judgements about the appropriate level and mechanisms for farmer contributions involves complex economic, financial, equity, political, administrative and legal considerations. Each of these considerations will require criteria or tests by which to judge policy options. These criteria will seldom if ever have equal weight nor always be consistent.

We conclude that rational decisions about changes in irrigation fees or other methods of resource mobilisation cannot be made without simultaneously reviewing the broader context of the nation's sectoral price and taxation policies and its general macroeconomic policies and environment. For this reason it is not possible to give simple, universal answers to questions such as 'what is the best approach to financing the recurrent costs of irrigation?' Rather, a framework of analysis must be developed, which can then be applied flexibly to individual situations. This is the challenge that we undertake in this book.

1.2.2 *Conceptual framework to analyse irrigation financing policies*

Our analysis of irrigation financing policies is guided primarily by the conceptual framework provided by the discipline of economics. We are economists by training, and we find the concepts of economics provide a useful framework for identifying both problems and policy options for irrigation.

But we also believe that the concepts of economics are too valuable to be left in the hands of economists! We therefore develop the essential concepts in ways that should be readily understandable to non-economists. At the same time, students of economics may benefit from the opportunity to consider how the fundamental concepts adorning their economics textbooks can be brought to life amid the realities of critical policy concerns for irrigation in the Third World.

One of the advantages of economics as a framework for policy analysis is that it provides two broad criteria, namely efficiency and equity, against which policies can be evaluated. Of the two, economics has tended to give greater emphasis to the efficiency criterion because it seems more 'objective' and therefore subject to more definitive conclusions. Equity, by contrast, is an inherently subjective concept, about which economists can seldom speak with authority. It is, however, a primary concern of those who call themselves 'political economists'.

In this book, we consider both efficiency and equity in evaluating irrigation financing policies. We identify four important efficiency criteria, each reflecting efficiency in the allocation or generation of one specific resource or set of resources. One general thesis underlying this book is that irrigation financing policies have the potential to affect, for better or for worse, the quality of performance of irrigation projects. The efficiency criteria that we use focus attention on the linkages between irrigation financing policies and irrigation performance.

But our book would be much too narrow if we limited our criteria to those derived from the concepts of economic efficiency. Irrigation financing policies involve decisions about who should pay how much for economic benefits provided to some as a result of public sector activities. Such decisions are inherently political in nature, and as such, they involve ideas (often conflicting ones) about equity. In the case of water, these conflicts often go deeper than with any other agricultural input with the possible exception of land. Fundamental attitudes about water often give the irrigation financing policy arena a highly charged emotional atmosphere. The importance of attitudes is well illustrated by the following quotation from a study on the Middle East:

> . . . the region's water resource quagmire is even deeper than technical, management, or economic constraints would suggest. More difficult to assess and alter are underlying passions. Although actual physical conditions vary from nation to nation, attitudes about water do not: in every country, access to clean water is considered an undeniable right, and tampering with water supplies is considered an unspeakable crime. Especially in more traditional agricultural areas, consumption patterns reflect deeply ingrained, age-old feelings about water. Water determines the nature of economic survival, permeates cultural norms, and infuses political ideology. Although technology may be harnessed, emotions pose the ultimate challenge.[3]

Equity questions can never be definitively answered by an external analyst. All that we can do is to identify equity as one of the criteria of financing policies, draw attention to the kinds of considerations that are relevant to include in a policy evaluation, and where necessary, point out inconsistencies or other inadequacies in the political statements relative to the relationships between equity and irrigation financing.

In addition to looking to economics for the analytic concepts that allow us to evaluate irrigation in terms of efficiency and equity criteria, we also need to incorporate institutional considerations into our evaluation of irrigation financing policies. In any given situation, financing policies and the economic forces acting upon them operate within a specific institutional setting consisting of such things as organisations, rules and laws, and administrative procedures. A nation's institutional setting reflects a variety of its social, economic, political, historical and cultural characteristics. We identify one key element of this institutional framework – the presence or absence of financial autonomy – that is of particular importance to an understanding of the likely performance of irrigation financing policies. This institutional factor turns out to be of major importance in evaluating financing policies with respect to most of the efficiency criteria.

1.2.3 *Testing conclusions against field experience*

This is not an 'armchair economics' book. If we had nothing further to say on irrigation financing policies than could be derived from economic theory and concepts, we would not have bothered to write this book. Over the past several years, we have tested the concepts and methods we have developed in field studies in a variety of countries. We draw liberally from these field experiences to give flesh and details to the points we make in this book. This book is not an abstract modelling exercise. Our concern, rather, is to devise methods for obtaining finance and allocating scarce resources to irrigation. We are thus engaged in a practical exercise in political economy.

1.2.4 *Looking ahead: a brief summary of the main arguments*

The combined effects of the expansion of irrigation over the past few decades and the fiscal crisis faced by many governments during the 1980s have brought increased attention on the shortcomings of policies for financing the provision of irrigation services. Particular emphasis is placed on the ways of financing the recurrent expenditures for operation and maintenance of facilities already built.

In today's atmosphere of 'get the prices right', many argue that user fees for irrigation water should be established or raised. While we also have a preference for user fees (and devote much of this book to an examination of various details about the operation of such fees) our preference is contingent on existence of financial autonomy for the irrigation agency. Under financial autonomy, a system of user fees has the potential (1) to improve irrigation operations both by freeing the O&M budget from the constraints imposed by the central government's fiscal difficulties, and by increasing the accountability of the irrigation system managers to the water users; and (2) to encourage a more appropriate and realistic evaluation of irrigation investment proposals. These potential efficiency benefits are lost in the absence of financial autonomy.

Many advocates of user fees assume that the fees will encourage farmers to be more efficient in their use of water. But the validity of this argument is contingent on the fee being structured in such a way that the farmer's total water bill will vary according to his water-use decisions. In reality, most systems of user fees in Third World countries are not structured in this manner. Rather, the fee is fixed on some basis related to the area farmed.

The debate over irrigation fees is also argued on equity grounds. Poor farmers, it is often stated, should not be made even poorer by imposing a user fee on them. We agree that there are situations where the overall policy and macroeconomic framework is so distorted and skewed against the rural sector that imposing user fees for irrigation would be inappropriate. This is less an argument against user fees that it is an argument in favour of placing top priority on creating a more balanced policy and economic environment for the farming sector. But what about the more typical situations? A careful look at the equity question will often reveal that (1) irrigation farmers are certainly poor, but (2) rain-fed farmers, landless labourers and many urban people are even poorer. User fees may thus serve equity even though they require payments from poor farmers.

In general, we feel it is not desirable to attempt to use irrigation financing policies to pursue broad goals of social equity and income redistribution. This is not to deny the importance of these goals; rather, it reflects two conclusions that we have reached: (1) that irrigation financing policy is a relatively ineffective tool for achieving these social goals; and (2) that efforts to use irrigation financing policies to do so severely reduce their ability to perform their primary task of financing irrigation services.

1.3 *Conclusion*

Irrigation services are seldom adequately financed. The economic potential for irrigation will be realised only if there are policy shifts, and resources for operations are mobilised much more reliably and widely than hitherto. We believe that irrigation farmers must assume a greater responsibility for providing finance, and that this is very much in their interest. This book examines the scope of mobilising resources with particular focus upon the role of farmers in this process.

However, devising workable mechanisms for mobilising this finance requires a deep understanding of agriculture production processes, and of the great variation in income both among farmers and, for each individual farmer, among years. It also requires an understanding and appreciation of the impact of key macroeconomic issues and the general policy framework. Above all those responsible for irrigation finance must have a keen political touch. Who should pay what, to finance public services, lies at the heart of political debate.

The funds for ensuring satisfactory irrigation development and maintenance will have to be raised from various public and private sources if the key opportunities for subsector development are to be realised. This book presents a detailed analysis of the economic, financial, social, political and administrative issues. We believe that economic concepts and methods present a useful framework and valid perspective for examining on these issues. We attempt to identify key economic concepts that will help in elaborating the central issues, in analysing options and in pinpointing preferred policies.

Students of economics will benefit from seeing a useful real world application of their concepts and methods instead of the unrealistic abstract and arid Robinson Crusoe world that characterises much economic teaching. General theoretical propositions can give a useful professional basis for analysis. Economic theory provides a valid framework for organising and analysing, and for considering problems and data, but there is no substitute for testing this theory with a study of practical problems. This is a major goal of this book.

PART I

Analysing financing policies: theory and concepts

Our analysis of financing policies is grounded in the conceptual framework provided by economic theory. In the following two chapters, we provide an overview of the key theoretical elements and concepts upon which our analysis is based.

Chapter 2, written particularly for non-economists and students of economics, introduces a few key concepts from neoclassical economic theory. We do not attempt a thorough and rigorous presentation of these concepts; rather our purpose is to identify the essence of their theoretical insights so as to make apparent their practical utility in the analysis of irrigation financing policies.

In Chapter 3 we develop a conceptual framework for examining irrigation financing policies. After characterising the myriad of financing methods into a few basic types, we use the concepts of economic efficiency and equity to identify five criteria for evaluating the desirability of alternative financing policies. Finally, we examine how a key institutional factor – the presence or absence of financial autonomy in the agencies responsible for operating the irrigation facilities – affects the likely outcomes of financing policies.

From the theory and concepts presented in these two chapters, we arrive at several conclusions relating to irrigation financing policies that colour much of the rest of the book. These conclusions are (1) that user fees implemented by financially autonomous irrigation agencies often have many advantages over other types of financing arrangements; (2) that most of the advantages of user fees are lost in the absence of financial autonomy; and (3) that there is no single 'best' financing policy for all situations.

It is because of the last of the above conclusions that this book cannot be in the form of a 'cookbook' that indicates precisely how to develop a sound policy for irrigation financing. Policy recommendations in any given situation need to be guided by detailed information about the specific circumstances of that situation. But they also need to be guided by a sound conceptual and theoretical framework, one that encourages the analyst to ask the right questions and to consider important alternatives. It is our goal in Part I to develop such a framework.

2

Key concepts from economic theory

There are numerous excellent textbooks of economics and many valuable industrial and agricultural case studies that use economic principles to explore investment and management options using the economist's powerful kitbag of concepts and methods. There have been pioneering studies of this type applied to the water sector in industrialised countries.[1] Irrigation policy studies that address the special problems of low income economies are less numerous.

This book tackles water resource development problems from the perspective of finance. Finance is clearly a crucial aspect of irrigation development and particularly intractable in poor countries. The thesis that underpins this text is:

- Irrigation is important and becoming crucial to food security, employment, and income growth in poor countries.
- Financial problems create shortages and uncertainty of supply of irrigation water that in turn inhibits efficient irrigation and deters complementary investments.
- Financial analysis and prescription is an exercise in political economy but economics provides a valuable conceptual framework to identify options and guide decision makers.
- The criteria or tests to guide action should incorporate these economic insights, albeit with a recognition of their limitations in real world situations.

Economists have a different perspective on technical problems from many professionals. They look carefully at objectives and harder and longer at alternatives, particularly the use of a little more and a little less of any valuable resource. They are concerned with the whole system and not just a part, and with alternative uses of the scarce resources.

Economists are thus concerned with trade-offs between resources in different uses, and trade-offs of resources in the same use now or in the future.

Many readers of this book will have a thorough grounding in economics. They can skip this chapter. The purpose is to introduce key concepts and selected theoretical insights to those non-economists that are to be dragged perhaps reluctantly through the political minefield that makes up water policy. Students of economics should stick with the chapter. There is nothing difficult here but it should show the practical utility of apparently arid and abstract elements of the discipline.

2.1 *Jargon: its use and abuse*

Some readers of this book will be, or wish to become, professional economists, financial analysts or accountants. Others will merely require sufficient awareness of the technicalities to assist understanding of the issues and to aid judgements about how to finance irrigation development. Economics provides a body of theory, principles and hypotheses that create a framework or logical structure for considering data provided by experts in other fields. The framework gives a basis for considering the implications of alternative ways of allocating scarce valuable resources amongst competing demands for their use, now or in the future. Using the concepts we can set criteria or explicit tests to aid judgements of options.

In any discipline jargon is created for two main purposes: the first is legitimate, the second is indefensible. The first is to encapsulate in a brief form a complicated concept or set of theoretical propositions. For example, to an economist the term elasticity conveys images in general terms of a measure of the degree of responsiveness of one variable to changes in another. Thus price elasticity of demand is the degree of responsiveness of the quantity demanded of a good, say irrigation water, to changes in its price. The term price elasticity is a useful shorthand to ease communication.

The second use of jargon is as a device to bemuse or mystify an audience as a warning to non-specialists not to trespass into a particular professional area. Often quite simple concepts are dressed up in jargon to prevent general access to all but those who are initiated, usually by passing examinations. It has to be said that many professional groups are all too often guilty of this type of intellectual arrogance.

In this chapter some key concepts are elaborated to aid later discussion. Anyone relying upon these explanations to impress colleagues or justify fat consultancy fees should beware. Only the essence of these concepts is conveyed. Readers are spared most of the pitfalls and qualifications. To aid student reflection the words that have a special technical connotation in economics are italicised in this chapter.

2.2 *Efficiency and equity*

Economists typically evaluate economic activities in terms of the two criteria of efficiency and equity. Efficiency is a difficult concept that can cause considerable confusion because it is used by different disciplines to mean different things. For example, engineers concerned with irrigation frequently speak of water-use efficiency, by which they mean the ratio of the amount of water used (say, by the crop plants) to the total amount of water supplied. The higher this ratio, the more 'efficient' the use of water.

But to an economist, maximising water-use efficiency is not likely to be efficient! Economists are concerned not only with water, but also with all the other inputs (labour, land, capital invested in irrigation facilities, managerial effort) that are used in the production of the crop. Economic efficiency occurs when all of the inputs are optimally balanced in all the production processes. Thus in simple terms, *economic efficiency* is the allocation of resources in ways that maximise their contribution to human well-being, within the constraints imposed by the existing distribution of wealth and income.[2]

Although efficiency is a useful concept, its focus is on producing a package of goods in an optimal fashion, while ignoring the question of how the goods, or the income associated with their production, is distributed. This brings us to the second broad criterion of economics – equity. Equity can be defined as 'fairness', and is not necessarily the same as 'equality'. Equity is thus a subjective concept that can be evaluated only against a subjective standard. Two commonly used standards of equity are: (1) equals should receive equal treatment (*horizontal equity*); and (2) income redistribution should be towards the poor (*vertical equity*).

Economists often have much less to say about equity than they do about efficiency. This is not because it is a less important concept, but rather because they lack any scientific or objective basis for determining the equity standards against which any particular policy or economic activity is to be judged. On the other hand, equity questions are critical to an understanding of the issues of political economy.

2.3 *Externalities*

Under certain conditions, prices can act as an effective method of communicating to individuals scattered throughout an economy how they should act to achieve economic efficiency. It is this presumption that prices can facilitate the achievement of economic efficiency in the allocation of resources that causes economists to place so much emphasis on prices.

But sometimes there are effects, either positive or negative, of an economic activity that are not reflected in prices. These effects are called *externalities*. For example, an irrigation project may lead to increases in certain vector-borne diseases. These diseases represent a real and potentially large negative externality of the irrigation project, since many of these negative health consequences are not reflected in the costs of producing irrigated crops. When externalities are present, market prices will not give the appropriate signals for socially optimal resource use.

2.4 *Diminishing returns and profit maximisation*

When water is used as an input in crop production, it contributes to the total volume, and thus value, of production. If a crop were provided with no irrigation water at all, the total production would be very low. If we provided a small amount of irrigation, production would be somewhat higher. As more and more water is added, production would increase even further until the maximum level of production was reached where water was not a limiting factor. The additional volume of production resulting from one more unit of water is called the *marginal product* of water, and the market value of this additional production is called the *value of the marginal product*. In almost all physical production processes, as the amount of one input is increased while all other inputs remain constant, the marginal product gradually decreases. This fact is known as the *law of diminishing returns*. Thus, as the amount of input increases, its marginal product, and therefore also the value of its marginal product, decreases.

Because of diminishing returns, it is generally not profitable to use an input to the point at which the maximum level of production is obtained. As a general rule, profit will be maximised when the farmer uses an input just up to the point where the value of the marginal product (i.e. the increase in gross returns resulting from the last additional unit of input) has dropped to the point where it just matches the additional cost (*marginal cost*) that the farmer must incur from using the input. At that point the increase in net returns from additional use of the input is zero.

In competitive markets the marginal cost of an input is its price; therefore profit is maximised when an input is used at the point where the value of its marginal product equals its price.

2.5 *Capital costs and recurrent costs*

Investments in irrigation require the expenditure of funds for the creation, operation, upkeep and occasional upgrading of the irrigation facilities. Economists and accountants often group these expenditures into two categories: capital costs and recurrent costs. *Capital costs* of irrigation are those costs associated with the initial construction, upgrading and major rehabilitation of the irrigation facilities. They are thus incurred at the time that an irrigation project is first constructed, and then again sporadically over the life of the project. *Recurrent costs*, on the other hand, are the annual costs of operating and maintaining the facilities. They are incurred more or less continuously over the life of a project.

Both capital and recurrent costs are part of the real economic costs of irrigation, so that when a proposed irrigation project is being evaluated from an economic perspective, the distinction between capital and recurrent costs is important only to the extent that the difference in the timing of the costs affects their present economic value (see Sections 2.11.2 and 2.11.3). But once a project has been built, the initial capital cost becomes a *sunk cost*, meaning that no future decisions can affect its magnitude. As a result, during much of the life of a project, decisions about the recurrent costs of irrigation – how much to spend for what specific activities – are the most important investment-related decisions because they will influence the productivity of the existing irrigation infrastructure.

2.6 *Demand and economic rent*

When economists talk about demand they generally mean *effective demand*, that is, demand backed up by money. In many areas of public policy in low-income countries there is great need but insufficient effective demand because of poverty. In public irrigation it is rare, for a whole set of reasons, for the full costs of the service to be charged for irrigation water supply. This does not mean that there is not effective demand for irrigation. It is very likely that farmers get their irrigation water at a much lower price than they would be willing to pay, given their perception of the value of irrigation water to their farming. In these circumstances, in economic jargon, farmers obtain *economic rents* from irrigation water.

Economic rent should not be confused with house rent, share-tenant rent and other general uses of the term. Economic rent is the income earned from the use of an input (water) in excess of its cost. Economic rent can be earned even in the case of an activity that is socially unprofitable if government subsidies relieve the farmer from the requirement to pay for part or all of the cost of the input.

This concept of untapped economic rent is at the heart of the main thrust of this book. Irrigation farmers are often poor but they are seldom the ultra-poor among rural people. We shall argue that in the irrigation subsector economic rents gained by irrigation farmers are often relatively high, that it is in the interest of economic efficiency and equity that rents be tapped by irrigation authorities, and that many of the management problems facing the irrigation industry will not be solved until and unless this tricky political economy problem is tackled.

2.6.1 *Demand for water*

To elaborate the issues we shall first have to discuss the nature of effective demand for a commodity such as water. We begin by noting a peculiar feature of water that has important practical implications for irrigation-financing policies. Water is very difficult to measure. Unlike many other agricultural inputs, it does not come in bags, boxes or other containers in which the quantity is clearly marked. Water flows, sometimes where it is supposed to and sometimes not; sometimes at a rapid rate and sometimes not; sometimes in the day and sometimes at night. Unless water use is measured either directly or by means of some reasonable proxy, it is not possible to establish a true price for it.

In many irrigation systems serving large numbers of small farmers, water prices do not exist because of the difficulties and costs of measurement. The lack of water prices, however, does not necessarily mean that the water is free. Farmers are often asked to pay an irrigation fee that depends on the size of their farms. But this fee becomes part of their fixed production costs, and no matter how high or low it is set, the effective price of the irrigation water (that is, its marginal cost to the water user) remains at zero.

Let us now consider the nature of the demand for irrigation water. A demand schedule measures the quantities of a good or service that would be purchased at varying prices, with other things held constant (e.g. prices of competing goods or complementary inputs such as fertilisers).

But if a price cannot be established for water, then the demand schedule has no meaning! It is important to remember in the following discussion that in many situations this is precisely the situation that prevails. Only in a relatively small number of irrigation systems do conditions currently permit the establishment of a true price for water.

In Fig. 2.1 we show in a simple model the irrigation demand schedule (*DD*) illustrating the amount of water demanded (say per season) at varying levels of water price. As increasing amounts of water are used, the law of diminishing returns causes the value to farmers of additional amounts of water (i.e. the value of the marginal product of water) to fall until eventually at Q_1 it reaches zero. If Q_2 were available farmers would be willing to pay P_1; conversely, if price P_1 were charged Q_2 would be demanded. If factors other than the price of water change, then the entire

Fig. 2.1. Demand schedule for irrigation water.

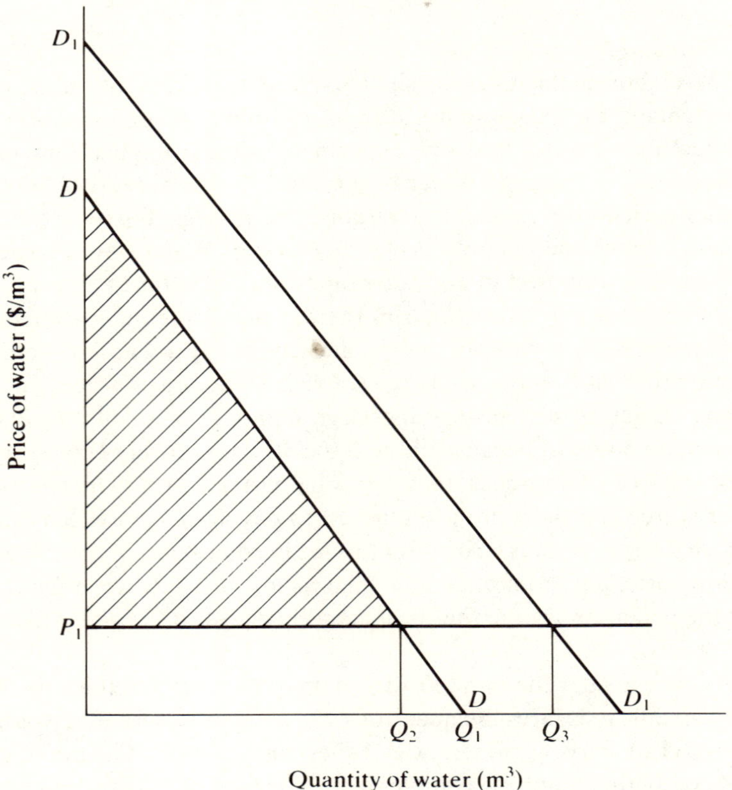

demand schedule will shift. For example, if new, more profitable technology was introduced (such as a new high-yielding crop variety or a valid extension message about timing of cultivation) then demand (and thus the demand curve) would shift to a higher level such as to D_1D_1. At this level of demand Q_3 of water would be required at price P_1. Any point on the demand schedule thus represents the willingness of the farmers to pay for water, which is equal to water's marginal value to them.

Returning to Fig. 2.1, if we focus for the moment on the original demand schedule DD we can see that farmers facing price P_1 will enjoy a surplus value equivalent to the shaded area. This is known as the *consumers surplus* in cost-benefit analysis literature but in this text we shall use the terminology *economic rent*. Economic rent is the excess of income earned by using an input (water, in this case) over its cost to the farmer (i.e. the sum of the values of the marginal product of each unit of water, minus the cost paid by the farmer).

We should perhaps note here that a taxing of the economic rents obtained by irrigation farmers can provide an incentive for them to market more produce to increase cash income, an obvious desirable side effect for governments keen to supply urban food demands from domestic sources. It can also play an important part in obtaining the full and proper use of agricultural land and the commercialisation of agriculture. The motives for holding land are seldom simply economic. There can be no doubt that in many countries landholders' political, social and longer term capital gains motives result in short-term underinvestment and neglect of irrigation facilities. It is often argued that much of the foundation for Japan's economic success was laid in the transformation of the rural economic structure from 1868 to about 1900, which was stimulated by the land tax. Tax and not subsidy can sometimes stimulate economic efficiency and growth.

In a typical public irrigation system the amount charged for water is not only low relative to the benefits from irrigation: it is low relative to the costs of providing water. In Fig. 2.2 we have superimposed the incremental cost of supply (curve SS), and the present capacity limit CC. Note that although costs fall at first (economies of scale) they rise quite steeply as diseconomies of size come in. The planners who installed CC misjudged effective demand, and there is no prospect of the scheme making financial profits unless there is an increase in effective demand, i.e. a shift of the demand curve to the right. The losses would be minimised if P_2 were charged unless some policy were brought in to enable the irrigation authorities to tap the economic rents (the large shaded triangle). For example a two-part tariff with a high connection fee for a base supply would reduce losses. If P_3 were charged for Q_3 and P_2 for any additional water then the total revenue would be $(P_3 \times Q_3) + (P_2 \times (Q_2 - Q_3))$, which is the same as $(P_2 \times Q_2) + ((P_3 - P_2) \times Q_3))$.

2.6.2 *Derived demand*

Irrigation facilities provide supplementary water to make crop production possible in deserts, or to increase yields in unreliable or inadequate rainfall areas. Irrigation water is an intermediate product used in the production of agricultural output. There is no direct demand for water coming from the consumers of the irrigated crops. The *demand* is *derived* from the profitability of crop production. It is a reflection of the value of the marginal product of water in crop production.

The intensity of demand for water (reflected in the height and the slope of the demand curve) is derived from the demand for the agricultural products. Changes in crop prices, or the introduction of new varieties

Fig. 2.2. Demand for irrigation water compared with marginal costs of its supply.

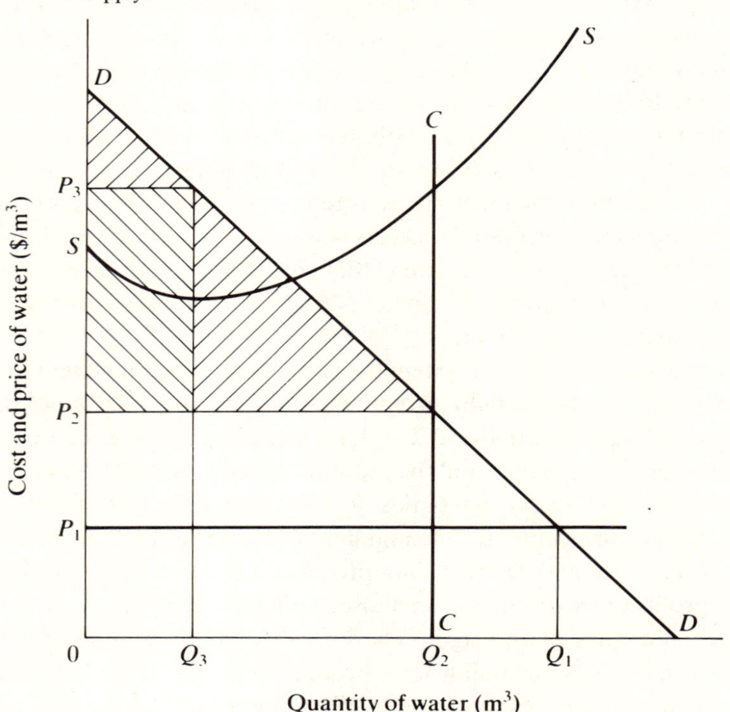

with greater yield responsiveness to water will cause farmers to shift their demand for water. It is therefore somewhat surprising that the profitability of irrigation is seldom at the fore in discussions of irrigation pricing policy.

2.7 *Irrigation financing and the role of the public sector*

Irrigation can be financed by public initiatives and/or by collective or individual private investment. A major reason why large-scale irrigation projects have been financed from public sources is because the scope of work was beyond private endeavour. Economies of scale in water resource development (falling average costs with increased size) often made the scope of work so massive that only government could command the resources necessary to get to the optimum level of investment.

Small projects such as an individual open well, or the joint construction by villagers of a furrow diverting water from a stream, may be within the means of individual or small groups of farmers. Many millions of such schemes are in existence; however, even they may be partially financed indirectly by public monies through government-backed credit schemes.

In his classic text on public finance, Musgrave[3] identifies three economic functions of government: the *allocation* of resources between public and private sectors; the *distribution* of income within the population; and the *stabilisation* of national income to avoid inflation or recession. These functions, in modern industrialised economies, expand the role of government well beyond Adam Smith's eighteenth century list of defence, administration of justice, and certain key public works. It is generally agreed there are several important functions that the private sector cannot achieve, and that in many countries, particularly low-income countries, there is a more active role for government which may extend through political choice to socialist or public rather than private activity.

Government is expected to promote growth, stability, equity (fairness) and economic efficiency. There are some defects in the way in which markets work that are relevant for certain aspects of irrigation policy that we shall study. We shall find that there may be a divergence between the private and social costs and benefits of irrigation. For example, if downstream effects of river development are positive (e.g. increased irrigation or hydropower potential as a result of upstream regulation) or negative (e.g. increased salinisation as a result of river disposal of saline drainage water) only government has the power to either tax gainers or compensate losers. If public fiscal policy squeezes agriculture's profits (e.g. by maintaining an overvalued exchange rate thus undervaluing agricultural exports) only government can make the necessary indemnity to recompense those who suffer to satisfy other government political goals. Government must provide (or ensure provision of) collective goods (discussed below) such as flood protection. In developing countries, where there is a general shortage of finance and a dearth of entrepreneurial talent, there will be a capital shortage and an aversion to risk-taking such as that inherent in large-scale water projects. In such circumstances it is not unusual for a more active set of public sector initiatives to be promoted. There may well also be legitimate political and ideological reasons for public sector investments that irrigation policy-makers must acknowledge.

2.7.1 *Financing public sector activities*

In the first instance, public sector activities may be financed from either external or internal sources. The primary external sources are foreign aid and international borrowing. External financing is frequently sought for new investments; it is generally not used to cover the recurrent costs associated with maintaining the investment, or with the operation of the completed facilities. Recurrent costs must generally be financed from internal sources. But even in the case of investments that are externally financed by borrowing, the eventual repayment necessitates that public sector activities ultimately be financed from internal sources.

A number of internal sources of finance for public sector activities exist. The use of general tax revenues as a source of financing is perhaps the most common approach. But public sector activities are sometimes financed by deficit spending. In this case, explicit taxes are not levied to raise the necessary funds; instead, an implicit tax, operating through the inflation which results from the deficit spending of the government, is imposed. User fees represent a third internal source of finance for public sector activities. This is often used when, in the case of investments such as irrigation facilities, there is a clearly identified group of direct beneficiaries of the investment who can be asked to pay for the services that the investment provides.

2.7.2 *Mechanisms for generating public sector revenue*

When people think of public sector revenues, they probably usually also think of taxation. *Taxation* is the compulsory transfer of money (sometimes goods and services) from private individuals, institutions or groups to the government. A tax can be directly levied on wealth or income or indirectly as a surcharge on prices. Deficit spending, which leads to general price inflation, is effectively another form of indirect taxation. Taxes thus represent a general obligation to pay for the costs of government services. The taxpayer's obligation arises because of his income or wealth or decision to consume a good on which taxes are levied; it does not arise because of any particular service that the individual has received.

But the public sector can also generate revenues through categories of mechanisms that relate payments in some way to the benefits of the service received.[4] Some of the most useful categories of these mechanisms are benefit taxes, special assessments, user fees, public prices and quasi-private prices.

- A *benefit tax* is a compulsory charge levied on people who are assumed to be the principal beneficiaries of a public investment or service such as irrigation or drainage. Little or no attempt is made to equate the level of the charge to either benefits or the amount of the service consumed. In some cases, as with urban landowners near irrigation projects, a benefit may exist even when none of the service is consumed.
- A *special assessment* may be considered as a type of benefit tax levied to defray the capital cost of improvements in the public interest such as regional drainage or flood protection. A betterment levy that shares the cost of irrigation between landowners and the government is a special assessment.
- A *user fee* is a compulsory charge levied on those who use a service to defray the cost of providing that service. It is assumed that the user receives a direct benefit from the service, and that the fee will be less than the benefit received by the user.
- A *quasi-private price* is a voluntary payment for a service sold by the public sector in the same way that a private firm would sell the service. The price fully covers the cost of providing the service, so that no subsidies exist. A fully charged public tubewell water supply would be an example of a service sold at a quasi-private price.
- A *public price* is similar to a quasi-private price, except that in setting the price, the government takes into consideration externalities that the purchaser cannot capture. In the case of positive externalities, a public price is therefore associated with a subsidy. If the government wishes to subsidise groundwater pumping to encourage drainage, then a 'public price' can be charged for public tubewell services.

2.8 *Collective goods and merit wants*

In allocating resources governments must take heed of the *social wants* for *collective goods*. Normally a consumer can be excluded from enjoyment of a good or service unless he or she pays for it. However, there is a group of goods and services where this cannot happen: if they are provided they can be enjoyed by all and it is not possible to exclude 'individual members' if they refuse to pay.

National defence is the classic collective good. People cannot be excluded from benefiting and it is indivisible – it cannot be sold individually but only to the community. The market mechanism is clearly defective with respect to the provision of collective goods. Private entrepreneurs will not provide the socially desired quantity of these goods because they could not expect to recoup their investment.

In the irrigation subsector, flood protection and regional drainage have some of these characteristics. Inasmuch as benefits are independent of the level of financial contribution, individuals have an incentive not to pay voluntarily. A farmer who does not pay receives the same benefit as one who pays. On the other hand, drainage and flood protection are not strictly collective goods. The amount of benefits that an individual receives is readily identifiable, and is largely in proportion to his or her landholding. An individual farm could be drained or protected, although the costs of so doing might make it prohibitively expensive.

Farmers are generally aware of the benefits of collective goods and services. They would vote for politicians who promise to provide them. But in some situations, the public may be unaware of the real benefits of an investment. In this case, a wise government may choose to promote the production of some goods above the level the market would dictate. Such goods are said to be *merit goods*, and to reflect *merit wants*. Merit goods are usually reserved for items such as health care or education services where consumers may not fully appreciate the potential benefits. In the irrigation field drainage is an area that might deserve merit good status. Farmers may not fully understand that the short- or long-term yield is depressed by the impact of waterlogging and salinity in the absence of costly drainage facilities. Another example arises in the provision of public health protection on irrigation schemes, where the risks of water-related diseases such as schistosomiasis and malaria are not fully understood by beneficiaries. For this reason government may be justified in providing a level of medical protection above that which farmers would 'purchase' for themselves. Where, in doing this, unhealthy swamps are drained or suchlike, there may also be an element of an external benefit, in that the general public also receives the benefit of reduced risks of infection. In these circumstances the farmers need not be expected to pay the full cost of the drainage as they only reap part of the direct benefit.

If it is simply left to farmers to purchase drainage or pay for health measures, or to put pressure through the political process for more public investment, the resulting level of investment is likely to be less than the social optimum. Judging by the general lack of drainage but increasing salinisation of arid-zone soils, and by the unhealthful conditions generated by many irrigation schemes, merit wants need to be given serious consideration.

2.9 *Public goods, private goods and public production*

Collective goods and merit goods are both *public goods*, reflecting *public wants*. In the former case the role of the government is to determine true consumer preferences that are not revealed by the market mechanism, and in the latter to actually 'interfere' with the consumers' own preferences. Although goods reflecting public wants must be paid for out of public funds, they may be produced by either the public or private sector.

Not all goods produced by the public sector reflect public wants. Often the public sector undertakes the production of many private goods (that is, goods that are neither collective goods nor merit goods), which may then be sold to individual citizens. To a considerable extent, the public provision of irrigation service falls into this category. It is the technological characteristics of water investments, in particular their falling average costs with size, that causes them to be produced by the public sector.

In the current economic and political climate there is much discussion about 'rolling back the boundaries of the State'. Experiments are being undertaken with privatisation of public investments, including water resource sector services. For example, public tubewells in India and Pakistan have been allocated to farmers and farmer groups. A wide variety of arrangements are possible. Public ownership does not have to mean public operation. Drainage wells could, for instance, be operated by contractors. Canals could be operated by water supply companies. One obvious potential advantage of such arrangements, to which we shall return later, is the separation of operation from monitoring and regulation of the standard of service.

2.10 *Elasticity concepts*

In Fig. 2.2 we can see that if price is increased from P_1 to P_2, the quantity demanded, if nothing else changes, declines from Q_1 to Q_2. The consumers are sensitive to the amount charged. We have already indicated that the price elasticity of demand is a measure of the degree of responsiveness of changes in quantity demanded (the dependent variable) to changes in price (the independent variable). In this case the price elasticity of demand would be obtained by measuring

$$\frac{\%\text{ change in quantity demanded}}{\%\text{ change in price}}$$

Readers will recognise that because quantity demanded and price change in opposite direction (as indicated by the slope of the demand schedule downward to the right), this elasticity is a negative number. The magnitude of the elasticity depends both on the slope of the demand schedule and on the level of price and quantity. Demand schedules are normally curves, convex to the origin, and calculus is used to calculate the elasticity at a point on the curve.

The position and shape of the demand schedule depends upon a number of features besides the price of the input. For example, in irrigation the demand for water will depend upon its price, as well as the price of output from irrigated crops, the price of complementary inputs such as fertiliser, the return to substitutes such as rain-fed farming, and the degree to which irrigation cost dominates the total budget.

The importance of a 'correct' price in attaining an economic allocation of a good depends in part on the nature of the good's elasticity of demand. If demand is very inelastic, a change in the price of the good will have little effect on the quantity that is purchased, and thus little effect on the economic efficiency of the use of the good.

Governments take this into account in choosing goods on which to levy an excise tax to generate government revenues. Goods such as tobacco and alcohol, for which the demand is relatively inelastic, are common targets for such taxes, since the price can be raised without much affecting the total volume of sales.

2.11 *Investment concepts*

Economics requires a broad view of investment. Often there are multiple, sometimes conflicting, objectives to be satisfied. Economic analysis helps to ascertain the contribution of alternative courses of action to these various objectives and to make explicit the trade-offs between them. This viewpoint is particularly relevant in determining sector allocations.

2.11.1 *Opportunity cost*

This is possibly the concept that the economist is most anxious that other planners should appreciate. The *opportunity cost* of devoting resources to a particular investment is the value of the best alternative opportunity that is thereby foregone. If resources are devoted to irrigation they cannot be used in another sector of the economy; if funds are invested in dams these same funds are not available for drainage; if resources (e.g. engineering expertise) are devoted to one scheme they are not available for another; if extra resources are devoted to projects that satisfy high design criteria, fewer people will be provided with irrigation water. The opportunity cost of an activity is thus a fundamental reflection of the activity's real cost to society.

2.11.2 *Time preference*

The value of a cost or a benefit varies depending upon when it is incurred. The further into the future that a cost or benefit occurs, the lower is the present value of that amount. This is justified on two grounds.

First, the total value of productive resources plus their output will increase over time, and therefore we can anticipate that the next generation will be better off and place a lower value or utility on their last unit of income. For this logic to be accepted we have to agree not only that average incomes will rise but that the value of say an additional 10 Rupees to a rich man is less – in terms of its utility to him – than the value that an additional 10 Rupees would have to a poor man.

Second, and less controversially, we can agree that in general individuals and society place a higher value on present consumption than on future consumption. Therefore adjustments must be made to the costs and benefits of a project to relate them all to a particular point in time, and weight is given to the present day with the weighting decreasing over time. Time preference is said to be positive.

2.11.3 *Discounting*

The means by which all costs and benefits are related to a particular point in time is discounting. The choice of the discount rate is important since the least cost solution for any given objective may change depending upon the choice of the discount rate. A low discount rate will favour high capital investment projects with low running costs. A higher discount rate will favour projects where the major proportion of costs are incurred in the future.

Basically there are two schools of thought on discount policy. There are those who hold that it is the responsibility of government not to discount the future too heavily. In fact governments have a unique responsibility to safeguard the future and to adjust the individual's 'defective telescopic faculty'. This requires a 'social time preference' rate which really reflects the politicians' and planners' weighting of present and future consumption.

The second school suggests that resources should be diverted to the projects that yield the greatest return. This leads to a high discount rate – a 'social opportunity cost' rate. High discount rates, 10% or more, are the 'norm' today. The effect of high rates of discount is to choose pump projects over gravity projects; thermal power over hydropower; cheap construction over substantial construction; projects that generate high recurrent costs over those with low recurrent costs. It discourages investment in catchment protection and forests and all projects that take a long time to come to fruition, no matter how big the ultimate benefit. It can provide an economic rationale for rapid mining of groundwater, deferring investment in drainage, for planting on hillsides and stream banks and literally for destroying the environment. High discount rates can give an economic rationale for hunting slow-maturing wildlife like elephants and whales to extinction.

In the area of irrigation finance the important impact is the tendency of high discount rates to increase the burden of recurrent costs in a project. This we shall return to because the trend is to try to get farmers to pay recurrent costs.

2.11.4 *Financial costs and shadow prices*

The financial cost of a resource is the cash that has to be paid to acquire its use. The economic cost is equivalent to the opportunity cost. Frequently the economic cost of an input is not equal to the financial cost that has to be paid. The most common examples cited in developing countries are the prices of unskilled labour and foreign exchange. The use on a project of an unemployed unskilled labourer may result in little or no loss of production elsewhere in the economy. The opportunity cost may be considered to be very low. (However, increased employment will lead to increased consumption of resources that have a positive opportunity cost. Therefore, even if there is no direct loss of output, opportunity cost will be greater than zero.)

Minimum wage rates, with statutory backing, are normally higher than opportunity costs. In contrast, the currencies of most developing countries are overvalued by the maintenance of unrealistic official exchange rates. Similarly, skilled labour is usually scarce and its financial cost does not fully reflect its value.

In an economic analysis the real value or opportunity cost of the resources to the economy must be used and, in cases where the financial cost does not reflect the economic value, corrections are made by means of *shadow prices*. A shadow price is thus simply an estimate of a good's true opportunity cost to society. Given the data available in developing countries, no precise calculations can be made of shadow prices, but an estimate in the right direction with an approximate order of magnitude will be a step forward. For example, it is suggested that in many countries with chronic unemployment, unskilled labour should be valued at between 25% and 50% of its wage or financial cost.

2.11.5 *Cost concepts and structure*

The *marginal cost* of production is the cost of delivering the last (or marginal) unit. The *average cost* per unit is the total cost of production divided by the number of units produced. Therefore, if marginal cost is less than average cost, average cost will be falling. Conversely, if it is greater than average cost, average cost will be rising. *Fixed costs* are constant as production varies, though the time and range must be specified. However, total *variable costs* alter as production alters.

The major feature of the cost structure of most irrigation supplies is that fixed capital costs are high and that there are significant economies of scale – that is, with increasing capacity, marginal costs are low and average costs are decreasing. For example, if pipe diameters are doubled, the water carrying capacity increases fivefold and the cost may approximately double. There may be high fixed costs for intakes, pumping plant and water distribution. In other words, much of the irrigation investment is 'lumpy'.

In addition to high fixed capital costs, a large proportion of the recurrent cost does not vary with consumption. Frequently the largest recurrent expenditure item is operation and maintenance staff which is usually independent of the level of water use. Hence, over a considerable operational range unit costs fall with an increasing level of capacity use.

2.12 *Application of key concepts to pricing of irrigation water*

Irrigation pricing should be designed to encourage the optimal use of water, from the standpoint of society. Optimal use is defined as the state in which society's overall level of welfare cannot be improved by reallocating water to other uses. Given a fixed quantity of water, the marginal social benefits of additional water allocated to each use should be equal. This will ensure that in all uses, the marginal value of water, or its opportunity cost, corresponds to the marginal social benefits in its best usage. If additional water can be obtained it will be at a marginal social cost in terms of other resources. Maximum social welfare is achieved when the marginal social cost of acquiring additional water for each use equals its marginal social benefits in each use.

The standard recommendation made by economists to achieve the optimal state is to set prices of inputs equal to marginal net social costs. These costs will include, at the least, the costs of producing and delivering water. But where the quantity of water is fixed (as in the case of an irrigation system based on a run-of-the-river diversion), the cost of 'producing' the water for delivery to one farmer is essentially the opportunity cost of having denied that water to other potential producers. As a result, a large component of the marginal social cost of water is likely to be in the form of the opportunity cost of the water, rather than in the form of explicit costs incurred by a public irrigation agency.

Ideally charges would be set at the marginal cost of water per unit volume (m^3), and the water would be sold volumetrically. A rational farmer would then use water until the marginal private benefit equals the price set at marginal costs. (Box 2.1 explains why this might not supply enough water for maximum evapotranspiration, and therefore why yield losses due to water shortage may not be eliminated.) Thus, the individual farmer's private decision would result in the socially optimal or economically efficient use of water across all users (provided that marginal private benefits were the same as marginal social benefits, i.e. that prices reflected social opportunity costs).

Several common problems exist with respect to charges for irrigation water. First, as noted in Section 2.6.1 on the demand for water, for a variety of reasons, including the problems associated with water measurement, there are difficulties in the establishment of true prices for irrigation water. Second, in the limited situations where prices do exist, the rates are generally well below costs. Additionally, rate-setting is seldom a purely economic activity, and non-economic factors often impinge upon the decision-making process.

BOX 2.1

Why maximum yield and maximum evapotranspiration are not optimum

In the planning for almost every irrigation project, the procedure used to determine installed capacity is to work out a technically feasible cropping pattern that meets predicted market opportunities. Crop physiologists then calculate the potential evapotranspiration of the crops and engineers design facilities to supply this amount of water. In arid zones an additional amount of water (typically 5–10%) is provided to leach the profile of salt.

Maximum evapotranspiration generally (but not always) produces maximum yield. However, with the exception of rice, the water-response functions for crops grown on farmers' fields generally have flat peaks such as shown in Fig. 2.3.

The figure shows that to reduce the water input from *X* (maximum evapotranspiration) to X^1 (by, say, 20%) will reduce yield, but only by less than 5%. If additional land and labour are available to make use of this water, the total physical returns to water on the project would be increased. The economic returns are also likely to be increased, depending upon the relative costs of alternative production strategies.

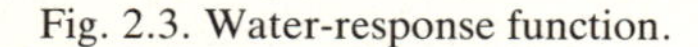

Fig. 2.3. Water-response function.

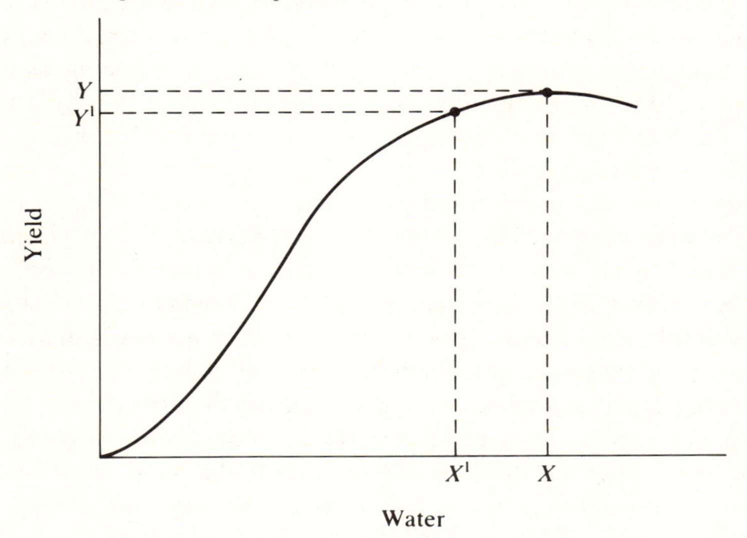

The diagram is contrived to make the point that maximum yield is not likely to be optimum. However, this reflects the underlying fundamental principle of diminishing marginal returns. It thus has practical validity and helps explain why farmers spread their water to increase the area cropped despite the normal extension advisers' urging to supply enough water to meet potential evapotranspiration. As a footnote we should add that irrigation capacity to meet leaching requirements is seldom needed as leaching can be carried out with existing capacity away from periods of peak demand.

The effect of designing for maximum yields is thus to increase installed capacity, and hence costs, above the economic optima.

2.13 *The market model and the difficulties of establishing water markets*

Based on certain assumptions and under idealised competitive conditions, it can be shown that the market system, through the interaction of producers and consumers acting in their own self-interest, can produce the optimal allocation of resources. Price is the regulating mechanism that influences market forces in the desired directions. When actual market conditions do not meet the idealised construct, regulatory processes or public production may be required to allocate resources. This may happen if externalities are present (as, for example, if there are uncompensated side effects such as pollution) or if there are economies of scale that set the preconditions for monopoly. These factors, along with others such as the relative plentitude of water in some areas, the physical characteristics of water, and the social values attached to it, explain the lack of well-developed water markets.

Among the physical attributes of water that prevent the full development of water markets are its mobility and its property of changing from solid to liquid to gas over the seasons. This makes it difficult to specify units of water, creates measurement problems and becomes an obstacle to the establishment of water (property) rights which are necessary for the exchange process in a market allocation system. Another constraint to measurement and the definition of rights is the fact that irrigation water is seldom fully 'used' by farmers. Most water users only consume a part of the available supply and the rest becomes accessible to downstream users. Measuring specific portions in this case is even more difficult.

The primary economic reason inhibiting the full development of water markets thus far has been the relative abundance of water compared with demand. Since water generally has not been scarce, except in certain places at specific times, formal institutions for managing scarcity (for example, markets) have not been created. However, it is important to remember that water is a widely different commodity in different places and at different times. For example, the value of water varies sharply depending on the seasons and upon crop needs. At some times of the year it is so valuable that it is known for farmers to fight and even kill for water; in the same society later in the year, or after a rain, excess water may be a problem and diverted to wasteland or downstream. As the demand for irrigation water grows and it becomes relatively more scarce, formal institutional arrangements for managing water will become increasingly important.

There is another reason why most societies have chosen non-market administrative means to allocate water. The demand for non-consumptive water use (such as recreational use) often represents a social want for a collective good, so that if market forces prevail, it is likely to result in an undersupply of the good.

Finally, there is a strong tradition of social values in many places that works against the adoption of market institutions for water allocation. Values that assign religious or ritual symbolic importance to water undermine tendencies toward market allocation by having a negative impact upon the individual's willingness to pay for water. Such values may be institutionalised in religious teachings that explicitly or implicitly prohibit market allocation for water. Politicians, always alive to opportunities to win in the popularity stakes, can undermine attempts to collect charges for irrigation water by popularising slogans such as 'water is free', a gift of God. We shall return to issues that this raises later in the book.

The overall impact of these factors is the general absence of developed water markets and the establishment of government rules and regulations for water use. Such rules seldom result in economically efficient resource allocation.

2.14 *Summary*

Advocates of market solutions to the problems of irrigation efficiency are confronted with several potential shortcomings:

(i) market imperfections exist that cause markets to function improperly;
(ii) markets give wrong signals by ignoring externalities;
(iii) markets do not work to provide the optimum level of public goods (including those for which public access cannot be denied;

those where public consumption does not deplete the benefits; and those where merit wants prevail); and

(iv) markets may yield undesirable results in terms of alternative objectives.

On the other hand, advocates of public intervention to correct market failure have to face the reality of bureaucratic failure in the allocation of resources. In developing countries administrative capacity is limited and often overburdened, and political leadership is at least as vulnerable as elsewhere to pressure groups (interest groups) who lobby for their narrow self-interest. There is certainly no guarantee that the public interest is served by the present political processes. Our observations of government failure and market failure lead us to advocate the solutions in this book that rely upon

(i) briefing of political leaders on the costs and benefits of policy options;

(ii) economic analysis of options;

(iii) decentralisation of authority for financing policy;

(iv) scheme level collection and retention of user fees, normally equivalent to operation and maintenance costs; and

(v) encouragement of collective user-group action to increase agency accountability.

3

Evaluating irrigation financing policies: a conceptual framework

To evaluate irrigation financing policies it is necessary to consider the objectives that such policies are designed to achieve. In this chapter we examine several possible objectives that policy-makers may have for irrigation financing policies. We then attempt to identify the types of conditions that must prevail if financing policies are to have the potential to achieve these objectives. In particular, we emphasise the conditions under which financing policies may be able to promote improved irrigation performance.

To begin our discussion, however, we need a framework for categorising the various financing methods that may be used as part of an irrigation financing policy.

3.1 *Methods of financing irrigation*

An irrigation financing policy is simply the combination of specific funding methods that a government or agency has established to pay for irrigation services. Although many different financing methods may be found in use by governments around the world, they can be grouped into a few basic types (Fig. 3.1).

The first distinction made in Fig. 3.1 is between direct and indirect financing methods. Direct financing methods require specified beneficiaries of irrigation to make payments linked either to the use of irrigation services, or to the benefits received from the existence of the irrigation facilities. In the former case, the payments are in the form of *user charges* (sometimes also termed *water charges*, *irrigation charges*, or *irrigation service fees*), while in the latter case they are in the form of a *benefit tax*. Indirect financing methods, by contrast, do not impose payments specifically for either the use or the benefits of irrigation.

Fig. 3.1. Irrigation financing methods.

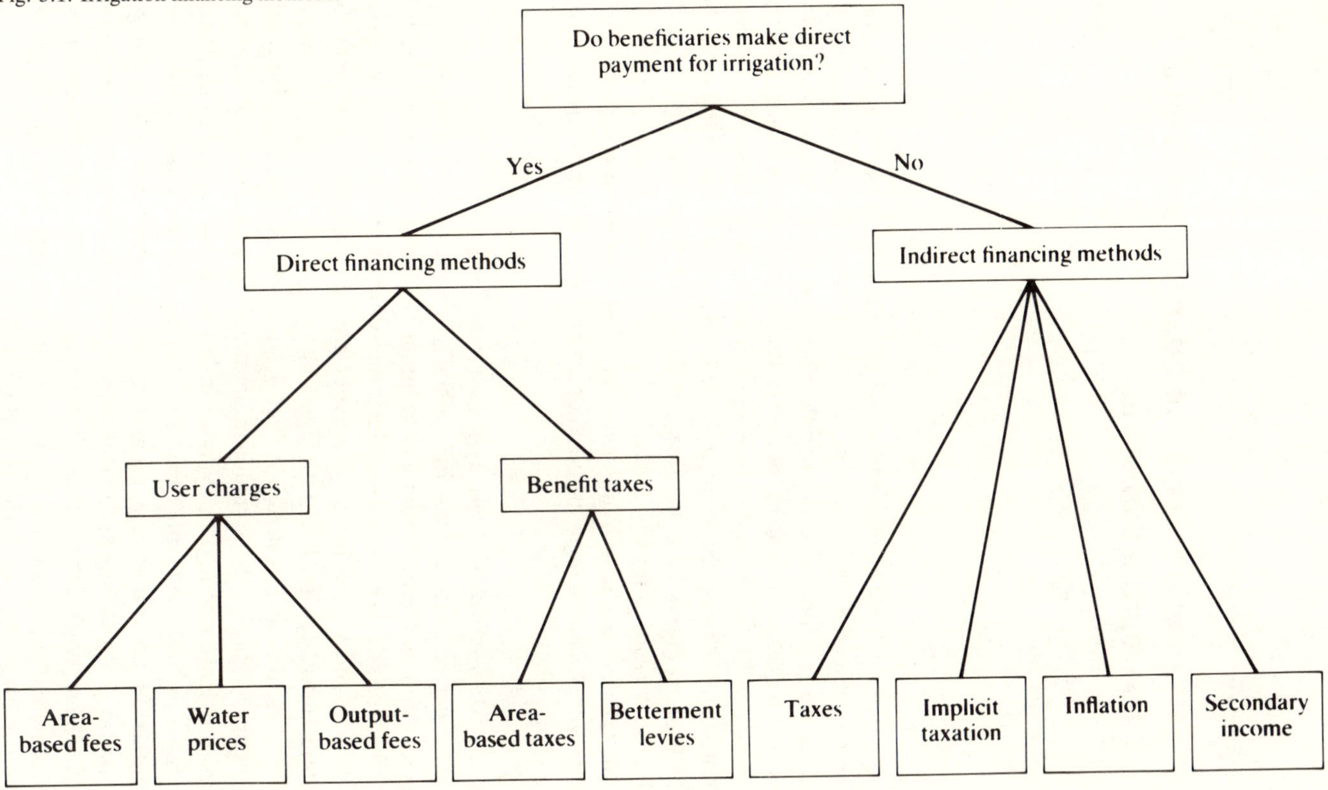

3.1.1 *Direct financing methods*

3.1.1.1 *User charges*

User charges are the most common direct method of irrigation financing. The many variants of user charges can be grouped into three primary categories according to the factors that affect the size of the charge (Fig. 3.1):

(i) area-based fees, with payments affected by cropping decisions made at the beginning of a crop season;
(ii) water prices, with payments affected by water-use decisions made during the crop season; and
(iii) output-based fees with payments affected by the level of production achieved at the end of the crop season.

In addition, some systems of user fees involve combinations of the above approaches.

Area-based fees: payments related to cropping decisions

User fees are frequently related to the cropped area served. The fee schedule may be very simple, with a single fixed amount per hectare cropped per season. But more complex fee schedules are also possible. For example, the fees per unit area may be differentiated by the cropping season ('season-wise rates'), by the type of crop grown ('crop-wise rates'), or both. Area-based fees are found in many countries including India, Pakistan and the Philippines.

In all of these cases, 'area' plays a direct role in the formula used to calculate the amount due. The amount of the fee may vary in accordance with a water user's decisions about the area, season and type of crop to irrigate; however, it does not vary according to the amount of water actually used.

All area-based fees share one key economic feature: throughout the cropping season, the fee for water is a fixed cost of production, regardless of the actual water-use decisions. In terms of the concept of demand discussed in Chapter 2, the effective price (or marginal cost) of water to the farmer is zero. As a result, the magnitude of the fee will have little effect on the quantity of water that farmers attempt to use.

Water prices: payments related to water-use decisions

Under a system of user charges that relates payments to water-use decisions, some measure (other than area) of the amount of water used is required. Possibilities include actual volumetric measurements to individual farmers; estimates of volume made on the basis of measurements at key points in the distribution system; and proxies for volume such as the length of time water is delivered, the size of offtake or depth over a weir, a pumping fuel bill or simply the number of times water is delivered. A system of water pricing requires that the farmers have some degree of control over the amount of water delivered. It is, therefore, primarily suited to systems that have an ability to allocate and deliver water in response to user demands.

Under water pricing systems, the total amount to be paid depends on the farmer's decisions about the amount of water to use. This causes the cost of water to become a variable rather than a fixed cost of production, thereby creating a financial incentive for the individual farmer to use less water than he or she would when the charge for water is a fixed cost.

Water prices based on volumetric measurements at the farm level are rarely used in large irrigation systems serving many small farmers. In part this reflects physical and administrative difficulties (and thus high cost) of controlling and measuring water volume to very small areas. Where we have come across meters in the field they have generally not been working.

It has sometimes been suggested that to overcome these problems, a system of volumetric water pricing could be instituted if the irrigation agency were to deliver and sell water in bulk at points higher in the system, where it would be more feasible to obtain reasonably accurate measurements of the volume of water delivered. Under such a 'water wholesaling' approach to user charges, the farmers served by each unit to which bulk delivery was made could be given responsibility for the ultimate distribution of water within the unit, and for collecting the funds needed to pay for the water received from the irrigation agency. China has reportedly experimented with this water wholesaling approach.

Output-based fees: payments related to production outcomes

In some situations irrigation service fees may be structured so that a water user's payment depends on the level of production achieved. For example, in areas where the entire irrigated area is devoted to a single crop, payment could be based on a percentage of total production. This type of fee structure is sometimes found in private irrigation systems. It is, of course, similar to crop-sharing arrangements for the rental of land. In the case of irrigation water, however, this approach has the advantage of giving the operator of the irrigation facilities an incentive to provide high-quality irrigation services. This type of fee structure is seldom used in government irrigation systems, although it has been reported in Vietnam.

Combinations of approaches: two-part charges

In some countries, the irrigation service fee is based on a two-part tariff, comprising a fixed or area-based charge and a variable charge. The fixed charge could make the irrigator eligible for some 'normal' or basic supply of water (which may be less than the amount typically desired), with the variable charge imposed on amounts of water taken in addition to this basic amount. Alternatively, the fixed charge could be in the form of an area-based capacity charge where each irrigator may contract for deliveries for a certain maximum area or for a maximum rate of flow. In this case a variable charge would then be imposed on the total amount of water actually consumed, measured on a volumetric basis. Such two-part charges are used in China and France.

3.1.1.2 *Benefit taxes*

Benefit taxes are a less widely used alternative to irrigation service fees. Unlike user charges, they are not necessarily linked to the use of irrigation water, or to the amount of irrigation service received. They are, however, linked in some fashion to the benefits received as a result of the existence of the irrigation system.

Area-based taxes

Assessments for irrigation are sometimes levied on the basis of the area commanded or served by the irrigation system, regardless of the type and number of irrigated crops produced in a year. Examples of this type of area-based tax can be found in Sri Lanka and Nepal.

Area-based taxes are superficially similar to area-based user fees; however, the amount of the tax to be paid does not necessarily bear any relationship to the use of irrigation. Payment is due simply because of the presumption that land lying within the command area of the irrigation system has benefited – either because it has actually been irrigated, or because it now has the potential to become irrigated. Like any land-based tax, the fee is a fixed cost to the farmer, unaffected by any cropping and water-use decisions.

Implicit in the concept of a uniform area-based tax is the idea that the benefits of irrigation are geographically distributed relatively uniformly throughout the command area. Although simple to administer, area-based taxes could lead to serious inequities in situations where the distribution of irrigation water gives some farmers much greater cropping opportunities than others.

Betterment levies

A betterment levy is a payment or a series of payments made by beneficiaries of irrigation specifically for the increase in the capital value of their land resulting from irrigation. The land need not necessarily be within the command area of an irrigation system. In cases where irrigation systems have stimulated urban development, betterment levies may be imposed on urban land under the presumption that urban land values have risen due to irrigation.

A betterment levy may be spread over a number of years, possibly with a grace period. For example, a charge could be equivalent to 30% of the increase in the capital value of the land, payable in three annual instalments, the first of which is due three years after the commencement of irrigation.

3.1.2 *Indirect financing methods*

Sometimes the public irrigation service is considered to be free. We have made the point already that in this case somebody somewhere is paying because the resources consumed in irrigation development and operation are scarce and have alternative uses. Furthermore, in cases where the government argues that water service is free, there are often a variety of informal charges, sometimes voluntary and sometimes compulsory and illegal, or levies of labour that operate at the individual farmer or user-group level. However, the main indirect method of financing is to make budget allocations to the agency from the government treasury. Government funds may be generated from a wide variety of types of *taxes*, such as land taxes, product taxes, sales taxes and general income taxes. Although in some cases the amount of a given type of tax that must be paid may increase as a result of the effects of irrigation, general taxes (unlike user fees and benefit taxes) are levied on individuals with no direct reference either to the use of or benefits received from irrigation.

Governments sometimes establish marketing boards that pay farmers artificially low prices for their products. This is particularly common in the case of export products. Generation of government funds through such arrangements involves *implicit taxation* of the farmers. Government funds may also result from deficit spending, in which case the financing will tend to result in *inflation*, which is in effect if not in reality another type of tax.

One other type of indirect method sometimes used to help finance irrigation is *secondary income* earned by an irrigation agency from sources other than irrigation service fees. These sources are not necessarily related to the primary functions of the agency. For example, an irrigation agency may be permitted to sell fishing rights on a reservoir used for storing irrigation water.

3.1.3 *Irrigation financing vs cost recovery*

A word of caution is in order at this point. Any of the five direct financing mechanisms discussed above and shown in Fig. 3.1 may be implemented in a country without being tied to irrigation financing. It is common, for example, in India and most ex-British colonies, that revenues from irrigation service fees become part of the general government revenues. They cannot be used to provide funds to operate and maintain irrigation facilities. The Treasury jealously guards its revenue collecting function and the rights of Parliament each year to allocate money to public purposes in the light of selected priorities. There is no 'earmarking' or reversing of income.

This leads us to make an important distinction between irrigation financing and irrigation cost recovery. Irrigation financing is the generation of funds that are specifically used to pay for the costs of providing irrigation services. Cost recovery, on the other hand, refers to the funds that flow into public agencies as a result of irrigation, regardless of whether or not these funds are used to pay for the costs of providing the irrigation services.

Like irrigation financing, cost recovery may be either direct or indirect. The above example of user fees in India involves direct cost recovery (or grant-in aid as it is sometimes quaintly called in South Asia). Water users make payments for irrigation whose magnitude is related to the amount of irrigation services received. But as the fees go into the general government revenues rather than being used to pay for the costs of the Irrigation Department, they are not a means of actually financing irrigation.

Indirect cost recovery mechanisms include a variety of general taxes whose collections rise as a result of irrigation. For example, revenues from Thailand's taxes on rice exports have increased with the increased volume of rice exports that have resulted in part from irrigation investments. The portion of the increase of these tax revenues that is attributable to irrigation is thus a form of irrigation cost recovery. But these monies are not earmarked to pay for the cost of operating Thailand's irrigation facilities. This tax is thus a means of irrigation cost recovery, but not of irrigation finance.

A common problem (and perhaps an understandable source of confusion) in many of the writings on irrigation financing policies is the failure to make this distinction between irrigation financing and irrigation cost recovery. Mechanisms, such as user fees in South Asia and the rice export taxes in Thailand, that increase the government's cost recovery from irrigation but which are not used to finance the costs of irrigation do nothing to improve the ability of the irrigation agency to operate and maintain the irrigation facilities in a satisfactory manner. The emphasis that we place in this book on the importance of user fees is predicated on their being actually used to pay for the costs of irrigation. Otherwise, they become, in effect, just another tax of dubious merit that rural people must pay. Economists in general would be concerned to see a high coincidence between any charges and finance. If taxpayers bear the costs there will be distorted signals to the farmers as to the real costs of the service. In addition, as discussed more fully below, there are real economic costs involved in raising taxation, a fact that is generally ignored in economic appraisals.

3.2 *Evaluating financing policies: What criteria should we use?*

Irrigation services, like all other goods and services produced in an economy, can only be provided by using scarce physical, financial and human resources. The fundamental role of irrigation financing policies is to assure the orderly acquisition of these scarce resources and their allocation at an appropriate level, form and timing to irrigation. From an economic perspective, these policies can be evaluated in terms of both economic efficiency and equity.

Although we are here, as in most of this book, concerned with an examination of irrigation financing policies from an economic perspective, we hasten to remind the reader that in any 'real world' situation, irrigation financing policies are subjected to evaluation from a political perspective. We explore some of these political factors in Chapter 12.

3.2.1 *Economic efficiency*

To evaluate fully the economic efficiency of irrigation financing policies, it is necessary to consider both their *resource-mobilisation efficiency* and their *resource-use efficiency*.[1]

Resource-mobilisation efficiency

Acquisition of resources (whether by a public or private entity) may entail costs of two types. *Administrative costs* (transaction costs) are cash outlays incurred specifically for the acquisition of resources. An example is the personnel costs of a tax collection agency. *Economic distortion* costs result when individuals and firms in the economy redirect their resources in ways that attempt to minimise the negative effects of taxes on their income or wealth. This reallocation inevitably creates distortions in the economy that lower the net value of the goods and services produced.

To evaluate an irrigation financing policy with respect to its resource-mobilisation efficiency, it is necessary to compare the total amount of resources generated as a result of the policy with the total opportunity costs (both administrative and economic distortion costs) incurred in the process of generating these resources. Other things being equal (which, of course, they seldom are in reality), irrigation financing policies with high resource-mobilisation efficiency would be preferable to those that mobilise resources in a less efficient manner. Unfortunately, all too often there is little or no information on either the administrative or economic distortion costs associated with financing policies.

Resource-use efficiency

Providing resources to irrigation is of course no guarantee that they will be used productively or efficiently. Efficiency in the use of resources is affected by many factors, some of which are largely independent of financing policies. But financing policies can create incentives that tend to affect the efficiency of the many resources used in irrigation. The possibilities for enhancing resource-use efficiency through financial policies thus need to be understood and evaluated.

Questions of resource-use efficiency arise at three stages in the irrigation process. The first stage involves the investment decisions associated with the planning, design and construction of the irrigation infrastructure. The second stage involves the operation and maintenance of the facilities once they have been built. This is often undertaken by a public or semi-public agency. Finally, resource-use efficiency questions arise with respect to the actual use of water by the farmers. An evaluation of financing policies thus needs to consider their effects in each of these three stages.

3.2.2 *Equity*

Irrigation financing policies affect the distribution of income among individuals and groups in the private sector of the economy – albeit generally in a modest way. Whether the effects are positive or negative, however, depends both on the policies and on the subjective view that one holds regarding an equitable distribution of income.

The fact that it is difficult, particularly for an outsider, to specify precisely what is equitable does not reduce the importance of equity as a criterion for evaluating irrigation financing policy. Any policy that requires users to pay for irrigation water needs to be seen by them as reasonably equitable; otherwise it is unlikely that the policy will gain the support needed to make it effective.

In summary, the concepts of economic efficiency and equity give us five important criteria for evaluating financing policies. These criteria are:

(i) resource-mobilisation efficiency;
(ii) quality of investment decisions;
(iii) cost-effectiveness of operation and maintenance;
(iv) water-use efficiency;
(v) equity.

3.3 *Evaluating financing policies: What can be expected?*

The extent to which any specific irrigation financing policy achieves the various results listed above can only be determined by a careful examination of the particular situation. But it is possible to make some generalisations about the conditions under which certain results are more likely to occur. To do so, we need to make an important distinction between *financially autonomous* irrigation agencies and those that are *centrally financed*.

Financial autonomy refers to an institutional arrangement whereby an irrigation agency must rely on direct financing methods for all or a significant portion of the resources it needs to operate and maintain the irrigation facilities. The agency also has operational control over the expenditure of these funds. Financial autonomy does not necessarily imply total financial self-sufficiency. In many cases governments allocate certain funds to financially autonomous irrigation agencies, particularly to cover a portion (often most or all) of the capital cost of irrigation development.

By contrast, with central financing, an irrigation agency's budget is completely or almost completely dependent on annual appropriations from a higher level of government. These annual budget allocations from the government thus determine the extent to which funds are available both for the construction of new projects and for the operation and maintenance of existing projects.

The distinction between financial autonomy and central financing is illustrated in Figs 3.2 and 3.3. Fig. 3.2 shows the relationships between irrigation expenditures, irrigation financing, and irrigation cost recovery in situations of financial autonomy. Except to the extent that foreign aid is used to fund irrigation expenditures, the amount of irrigation expenditures must equal the total amount of (domestic) irrigation financing. Funds from direct cost recovery are typically less than this amount, with the remaining financing either from funds allocated through government budgets or from the secondary income of the irrigation agency. The two sets of double arrows linking irrigation financing with both irrigation expenditures and direct cost recovery reflect the fact that the amount of

Fig. 3.2. Irrigation financing and cost recovery: financial autonomy (financing linked to direct cost recovery).

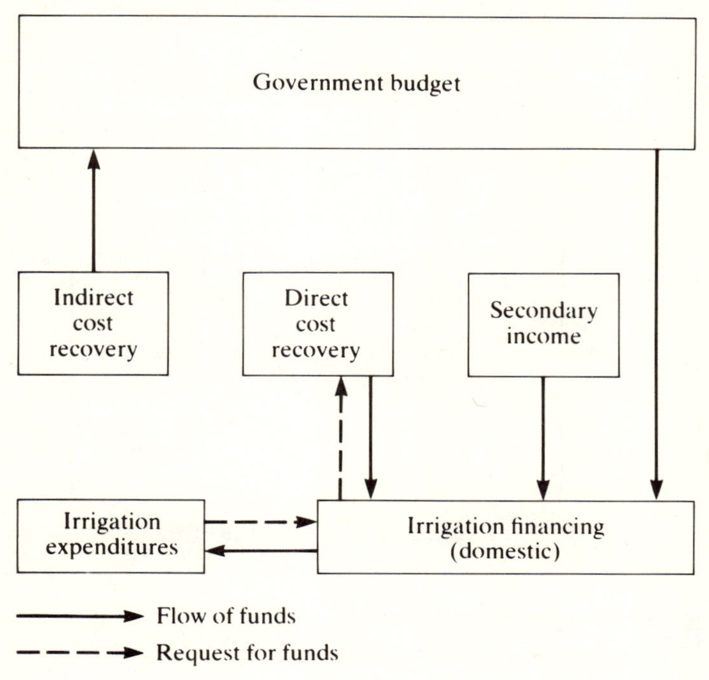

funds is jointly determined by an interaction between those making expenditure decisions and those making the financing decisions.

Fig. 3.3 shows the relationships between irrigation expenditures, irrigation financing, and irrigation cost recovery that typically prevail in situations of central financing. Again the amount of expenditures equals the amount of financing, but the only source of financing is the government budget. The interaction between financing decisions and expenditure decisions is now confined to the government budgetary process. Whatever funds are generated by irrigation cost recovery flow to the general government treasury and have no direct bearing on the amount of funds allocated to finance irrigation expenditures. Any shortfall in funds is likely to be made up by substandard maintenance and sometimes 'voluntary' donations from the farmers.

This distinction between financial autonomy and central financing is helpful in evaluating the likely effects of specific financing policies

Fig. 3.3. Irrigation financing and cost recovery: central financing (financing separated from cost recovery).

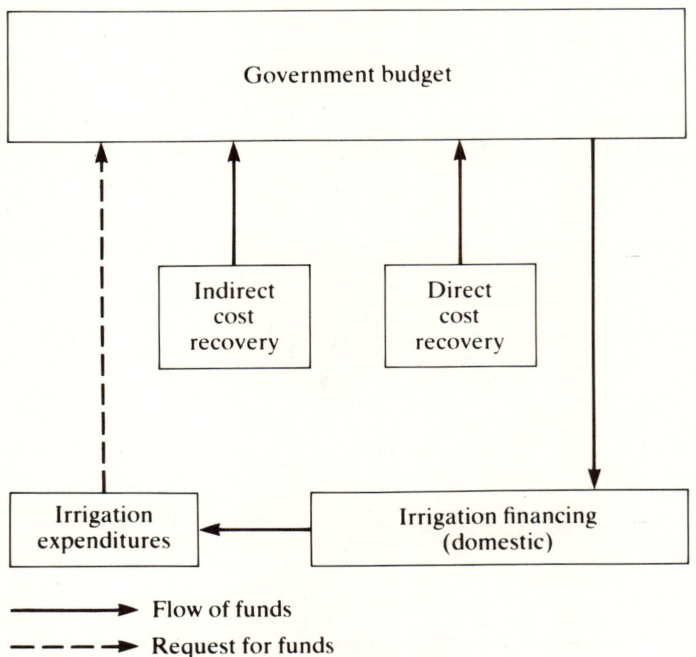

relative to the five criteria identified in the previous section. These effects are discussed below, and are summarised in Table 3.1.

3.3.1 *Resource-mobilisation efficiency*

A government decision to invest in an irrigation project ought to imply a commitment to the domestic financing of both the operation and maintenance (O&M) costs and all the investment costs except those covered by foreign grant aid. (Investment costs initially financed by foreign loans must presumably eventually be repaid! Even with foreign aid, a grant for one sector may preclude a grant to another, thereby creating an opportunity cost.) Whether this domestic financing is undertaken from general tax revenues, from special taxes or levies, from direct cost recovery measures, or from deficit spending (inflationary financing), there is inevitably a cost to the economy associated with the government's acquisition of these funds. As noted above, these costs include both direct administrative costs and the less visible opportunity costs associated with the economic distortions created by the taxation method.

Table 3.1. *Criteria for evaluating financing policies and potential effects*

Criterion for evaluation	Financial policy condition conducive to creating potential for favourable effect	Type of financing measures consistent with condition in column 2
Resource-mobilisation efficiency	Proposed financing measures can be implemented at lower total cost to society than alternative measures	Direct or indirect
Quality of investment decisions	Those with a financial stake in the success of the investment make the investment decisions	Direct
Cost-effectiveness of operation and maintenance	Financially autonomous irrigation agency	Direct
Water use efficiency		
via crop decisions	Water user's payment depends on crop decisions	Direct
via water-use decisions	Water user's payment depends on water-use decisions	Water pricing
Equity	Financing measures permit distinctions among groups or individuals perceived to deserve differential treatment	Direct or indirect

The effect of any particular financing method on resource mobilisation efficiency is thus an empirical question that depends on the method's costs of acquiring the needed funds relative to the costs of alternative financing methods. The question of resource-mobilisation efficiency is explored in greater depth in Chapter 7.

3.3.2 *Quality of investment decisions*

Good investments, whether in new projects or in full or partial rehabilitation or even small improvements to existing projects, depend on good investment decisions. In most countries, rich and poor, landowners are able to exert effective political pressure for irrigation investments. Direct financing or cost recovery methods that require the landowners to bear a significant portion of the investment and subsequent O&M costs may reduce their pressures for investments that are of dubious economic value to the nation. Inherent in the practical discipline of investment appraisal should be a determination to give the direct beneficiaries appropriate signals regarding the real (opportunity) costs of the resources to be consumed.

Where financial autonomy exists, the irrigation agency could have financial responsibility for the repayment of a specified portion of the investment costs. The agency in turn would have to obtain these funds from payments that it would collect from irrigation users. Under this financing arrangement, both the farmers and the irrigation agency would have a financial incentive to appraise, in advance, the economic viability of any proposed investment.

3.3.3 *Cost-effectiveness of operation and maintenance*

Promoting conditions that improve the operation and maintenance of irrigation projects is a potentially important benefit of a well-designed and implemented irrigation financing policy. When direct financing methods are used in conjunction with financial autonomy, three conditions are created that have the potential to enhance O&M. First of all, the budget available for O&M will be affected by the amount collected from the water users. This can act as an incentive for the irrigation agency to collect fees from the water users. Second, incentives arise for increased accountability of the irrigation system managers to the water users. The managers of the system, knowing that they must depend on funds collected from the users, are more likely to be concerned about the quality of the irrigation services provided. We believe that this enhanced accountability may, in the long run, prove to be the most important contribution of irrigation financing to improved irrigation performance. Finally, financial autonomy may create incentives for increased involvement of the water users in O&M. By undertaking certain activities themselves, the water users may be able to lower the fees that they would otherwise have to pay.

3.3.4 *Water-use efficiency*

In many cases where irrigation water is used inefficiently, a considerable part of the problem can be traced to poor operation of the main irrigation system. For example, the irrigation staff are in short supply or are poorly trained or motivated, water is not delivered on time and is not turned off after a rain, control structures are missing or damaged, and so forth. Thus, to the extent that financing measures are successful in improving the quality of O&M of the irrigation system, an increase in the efficiency of water use can be expected.

It is frequently argued, however, that financing or cost recovery measures will directly cause the farmers to be more efficient in their use of water. For this to occur, these measures need to affect either their *water-use decisions*, their *cropping decisions*, or both. Whether this is likely to happen depends on the type of financing method used. Financing methods cannot be expected to affect water-use decisions unless true water pricing is used and even then only if the water price is high enough to have a significant effect on the total cost of production.

Cropping decisions, on the other hand, could in principle be affected either by water pricing or by area-based fees that are differentiated according to the crop grown. For example, in many parts of Asia, farmers prefer to grow irrigated rice, even though wetland rice production generally uses much more water than would be needed for other crops. A higher irrigation fee for producing one hectare of rice than for one hectare of some other crop might encourage farmers to choose the crop that used less water. Frequently, however, the magnitude of the differential between rice and other crops is too small to be a significant factor in the cropping decision. An example from Pakistan is given in Box 3.1. In Egypt, a recent study also concluded that a system of charges differentiated by crop would be unlikely to have any significant effect on cropping decisions.[2]

3.3.5 *Equity*

Irrigation financing mechanisms can have an effect (albeit generally a very modest one) on the distribution of income among individuals

BOX 3.1

Using irrigation fees to influence cropping decisions: a hypothetical example from Pakistan[3]

In the Punjab in Pakistan, area-based irrigation fees are differentiated according to the crop planted. In 1985–86, irrigation fees on annual crops (in areas not served by Salinity Control Tubewells) ranged from Rs 11.2 per acre for *rabi* (winter) fodder to Rs 33.6 per acre for cotton. (Fees for other major crops were Rs 32 for rice, Rs 23 for oilseeds, Rs 21.6 for wheat, Rs 19.2 for maize and Rs 13.6 for *kharif* (summer) fodder.)

In Table 3.2, these seven crops are ranked according to their reported profitability, as measured by their net cash returns (column 4). In order to make the calculations, all irrigation charges were ignored, and the profitability of each crop relative to that of the immediately next lower-ranked crop was calculated (column 5).

With the help of a few hypothetical calculations, we can use the data in Table 3.2 to illustrate the difficulty of influencing cropping patterns by modifying the relative irrigation fee for any particular crop. For any particular crop, we make the hypothetical assumption that if the government wanted to discourage its production, the irrigation fee for that crop would be set at the maximum level consistent with the 1985–86 fee structure (Rs 33.6 per acre), and the fee on the next most profitable crop would be set at the minimum level (Rs 11.2 per acre). Based on these assumptions, calculations were made of the profitability of each crop relative to the next lower-ranked crop (column 6 of Table 3.2).

Table 3.2. *Effects of differential irrigation fees on relative profitability of crops: an example from Punjab Province, Pakistan*

				Profitability index relative to next lower ranking crop	
Crop (1)	Profitability ranking (2)	Gross return (Rs/acre) (3)	Net cash returns (Rs/acre) (4)	Assuming no irrigation fee (5)	Assuming maximum irrigation fee differential (6)
Rabi fodder	1	2407	1807	+12.9	+11.6
Rice	2	2519	1601	+0.5	−0.9
Cotton	3	2614	1593	+9.6	+8.2
Wheat	4	2231	1453	+19.2	+17.5
Oilseed	5	1720	1219	+5.8	+3.9
Maize	6	1811	1152	+15.5	+13.5
Kharif fodder	7	1344	997	–	–

The differences between column 5 and 6 represent the maximum effects on relative profitability that could be induced by manipulation of the irrigation fees within their existing range. A comparison of these two columns shows that for all crops, the change in relative profitability is small. Only in the case of rice, which appears marginally more profitable than cotton in the absence of any irrigation fee and marginally less profitable with the hypothetical fee structure, did the profitability rankings of the crops change.

These data suggest that it is quite unlikely that setting different water charges for different crops will, within the context of the general level of charges prevailing in Pakistan, have any significant effect on the cropping decisions of the farmers.

---

and groups in the private sector. Whether any particular mechanism has a positive or negative effect on equity depends both on the nature of the mechanism and on the subjective view that one holds regarding equity.

Two commonly identified concerns relating to the equity objective are differences in income among project beneficiaries, and differences between project beneficiaries and rain-fed farmers. Regarding the first of these, it is sometimes suggested that financing measures should be structured to account for differences in income levels among the project beneficiaries. Beneficiaries with higher incomes would thus be charged a greater amount, relative to the irrigation services or benefits received, than those with lower incomes. Very poor consumers of water could be exempt. One problem with this approach is that it tends to add considerable administrative complexity to the financing process. Thus the effort to increase equity may impose an unreasonable cost in terms of reduced resource-mobilisation efficiency. Furthermore, research shows that this type of 'targeting' generally misses the target, resulting in subsidies flowing to the wrong group.

Regarding the second concern, there is a presumption that payment by irrigated farmers for the high costs of irrigation is equitable because irrigation has made them better off than equivalent rain-fed farmers who farm without irrigation. This increase in income potential is often reflected in a rise in land values.

Although high irrigated farm income may occur in most cases, it would be wrong to assume it is always the case. For example, rain-fed farms may be larger than irrigated farms, so that income per hectare is not a valid comparison. Or, in situations where the irrigation project involves settlement of people in previously sparsely inhabited areas, irrigated farmers will have real and psychological dislocation costs and may indeed face lower incomes for many years than their longer-established counterparts in rain-fed areas. Where farmers are resettled from risky but high-value rain-fed crops to, say, rice farms, a drop in average income may occur. All this indicates that farm income data, with and without irrigation, and for both good and bad years is essential for policy analysis.

Our approach to equity issues is to try to separate them from water charges. We do not believe that user fees for irrigation water are a good basis for income redistribution. There is a case for providing basic services free or at a subsidised rate to low income people but this is a social welfare matter and requires a special programme. Subsidising irrigation water is no better a means for income support than subsidising any other input such as fertiliser or seed, or than giving farmers premium prices for their production. We prefer not to saddle irrigation agencies, nor any other input supply agency, with the task of administering income support programmes.

3.4 *Summary*

Irrigation financing must be distinguished from cost recovery. We are concerned with financing because of its focus on the provision of the financial resources for actually providing irrigation services.

Irrigation financing policies must be evaluated against criteria that reflect the objectives that policy-makers have established. From the broad criteria of economic efficiency and equity, five specific criteria of policy for irrigation can be derived: resource-mobilisation efficiency; quality of investment decisions; cost-effectiveness of operation and maintenance; water-use efficiency; and equity.

Resource-mobilisation efficiency is concerned with minimising the social costs of acquiring the resources to finance irrigation services. The resource-mobilisation efficiency of any particular financing method is an empirical question that must be evaluated within each individual nation.

Many of the potential effects of financing policies depend on whether the irrigation agency is financially autonomous, or whether it is centrally financed by the government. Policies that involve user fees implemented by financially autonomous irrigation agencies have the potential both to improve investment decisions and to encourage more effective operation and maintenance of the irrigation facilities. But user fees implemented by centrally financed irrigation agencies are unlikely to have any such beneficial effects. For this reason we place considerable emphasis throughout this book on the importance of financial autonomy.

User fees can encourage efficient use of water by the farmers, but only if they are in the form of true water prices. A variety of implementation difficulties, including problems of measuring the amounts of water delivered to large numbers of small farmers make the use of water prices relatively rare, except where water is very scarce.

Finally, equity is an extremely important, albeit subjective, criterion against which to evaluate financial policies. The chief concern should be that the water users perceive the financing policies to be equitable. In our view, a system of user fees can be and normally is quite consistent with equity goals.

PART II

Criteria for evaluating irrigation financing policies

Irrigation financing policies can be deemed successful to the extent that they achieve desirable objectives. In Chapter 3 of Part I, we identified five criteria reflecting alternative objectives for financing policies, and against which they could be evaluated. In Part II we examine in some detail the policy issues associated with each of these objectives.

The first three chapters of Part II deal with objectives related to the efficiency with which the various resources devoted to irrigation are used. We begin in Chapter 4 by examining to what extent, and under what conditions, financing policies can be expected to promote effective use of the resources devoted to operation and maintenance. It is our contention that although the link may not be direct, inappropriate financial policies are an often-overlooked contributing factor to the common problems of poor irrigation operation and maintenance. In Chapter 5 we turn our attention to the relationships between financial policies and the efficiency with which water itself is used by the farmers. Finally, in Chapter 6 we consider how financial policies affect the decisions that commit resources to irrigation through the investment process.

In Chapter 7 we turn to an examination of a different type of efficiency objective, namely, the efficiency of the process by which the resources needed for irrigation are acquired.

Efficiency is but one of two broad objectives commonly identified by economics. The other is equity, which is the topic of Chapter 8. Economists are often somewhat uncomfortable dealing with equity questions because of their inherent subjectivity. Yet such questions are probably more critical than are efficiency concerns in determining the overall acceptability of irrigation financing policies.

Throughout Part II we emphasise the key role that financial autonomy plays relative to most of these objectives. Our basic argument is that within the context of financial autonomy, systems of user fees for irrigation have the potential to further many of these objectives.

4

Cost-effective operation and maintenance

4.1 *Introduction*

Governments in many nations have invested large amounts of money in irrigation projects. More than $500 billion of capital has been spent. This is on the expectation that such investments would repay their costs and more generally foster economic growth and development. Realisation of this expectation clearly requires that projects be operated in a reasonably effective manner, and this in turn requires that an appropriate schedule of maintenance is followed. But irrigation operation and maintenance (O&M) are frequently reported to be highly inadequate, and this is thought to be a major contributing cause to the disappointing levels of agricultural productivity. This substandard O&M will depress economic returns to many irrigation investments.

These problems with O&M suggest two important policy issues with respect to irrigation financing: how to obtain finance and how to utilise the funds wisely. The relationships between these issues are illustrated in Fig. 4.1. The concern with obtaining finance is indicated in the upper left-hand portion of the figure, which shows funding decisions determining the amount of funds available for O&M. The issue of the utilisation of the funds involves consideration both of expenditure decisions that provide the irrigation agency with the physical resources to undertake O&M, and resource-utilisation decisions regarding the deployment of these resources in O&M activities. Fig. 4.1 illustrates the importance of all three types of decisions (funding, expenditure and resource-utilisation) on the ultimate effectiveness of O&M.

The first policy issue (how to obtain finance) is a budgetary issue commonly expressed as the problem of *how to provide the funds required for O&M*. Formulation of the issue in this manner, however, reflects a technological bias, in that it implicitly assumes the existence of some objective technological standard that determines the proper form and intensity of O&M activities. But it is reasonable to expect that the fundamental economic principle of diminishing returns will apply to O&M activities as it does to other resource allocation problems.

For example, a hypothetical irrigation project might be operated and maintained in a reasonably satisfactory way with an annual budget of, say $2 million. Some improvement in irrigation performance could be expected if $1 million more were added to the O&M budget. But if the additional returns to the economy from this increased or marginal expenditure were to be only $0.4 million, it would clearly be economically unjustified to increase the budget to improve O&M. The marginal return would be less than the marginal cost.

If, on the other hand, the marginal returns in the above example were $1.4 million, then a necessary condition for the expenditure to be economically justified is met – namely, that returns must be greater than expenditures. However, to be sure that such an expenditure is an

Fig. 4.1. Relationships between funding and effectiveness of O&M.

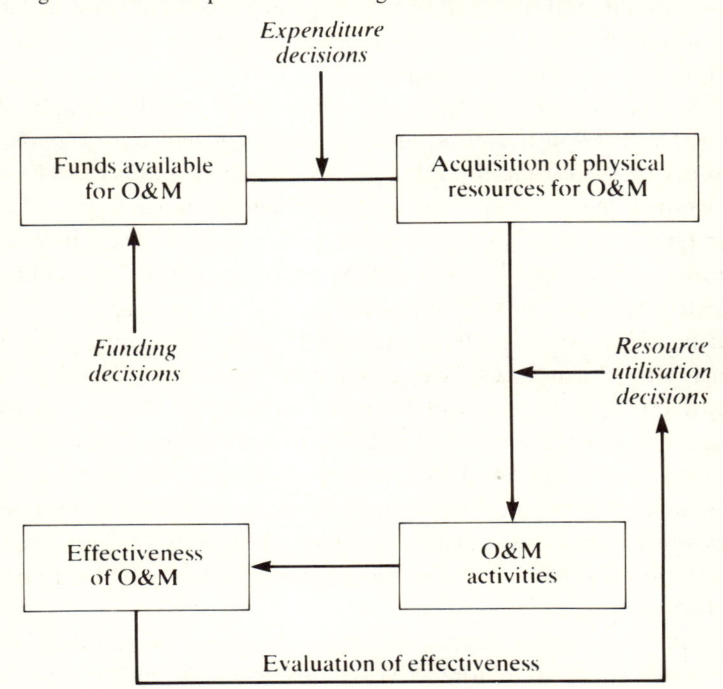

economically justifiable use of public funds, we must establish that this return is greater than the expected return from spending the money elsewhere – for instance, on electricity supply or road maintenance. In most developing countries severely limited public revenues make the opportunity cost of public expenditures very high.

This helps explain why funding requests based on the apparently self-evident technical case for maintenance are sometimes turned down by Treasury who, with limited finance, must consider trade-offs among many desirable budget requests. A general shortage of government funds may require that achievement of the economically optimal position of equi-marginal returns results in costly losses in terms of irrigation benefits foregone. Where government funds are extremely scarce a policy of equi-misery for all is quite rational. This suggests, however, that if the farmers in the hypothetical project of the above example could ensure that their irrigation fees would be used for the project's O&M, it would be in their interests to provide more funding for O&M than the optimal amount that would be provided by Treasury.

We therefore need to rephrase the first policy issue raised above as that of *how to provide the appropriate level of funds for O&M*. Obviously, this raises the difficult question of how to determine what level is 'appropriate' or optimal. As the following paragraphs make clear, there exists no unique answer to this question.

The second issue (one that tends not to be as readily recognised as the first) is that of *how to assure that the resources provided for O&M are used effectively*. Increasing the amount of funds available for O&M will not necessarily lead to improved irrigation performance if poor financial, managerial and operational decisions and procedures prevail. We need to consider, therefore, to what extent financial policies can enhance the effectiveness with which the resources are deployed.

But we cannot talk about the effectiveness of resource use without considering from whose viewpoint effectiveness is considered. What seems effective to a design engineer working in the capital city of a country may seem unnecessarily lavish to a project manager and wasteful or possibly even harmful to a farmer working in the fields. If financial policies are to play a role in enhancing the effectiveness of resource use, they must encourage a process of decision-making that involves evaluation and feedback by all categories of people concerned with irrigation performance.

Ultimately, this decision-making process implies a joint determination of how to use resources for O&M and how much should be spent. We thus arrive back at the first issue and will find that those involved in irrigation must evaluate collectively the alternatives available to them to determine what is appropriate O&M and how much should be spent to achieve it. Thus, as suggested by the chapter's title, the two primary issues discussed in this chapter – *cost* and *effectiveness* of O&M – need to be intimately linked together. To the extent that financing policies foster this linkage (the 'missing link' in Fig. 4.1 between effectiveness of O&M and funding decisions) they create the potential for better irrigation performance through improved O&M.

4.2 *Providing the appropriate level of funds for O&M: the budget process*

4.2.1 *Funding O&M in the context of central financing*

Most irrigation agencies operate within the context of central financing, dependent for their budget on central government allocations. In these circumstances obtaining an adequate budget for O&M is a difficult problem. The overall fiscal situation of many governments has worsened in recent years and therefore the problem of inadequate budgets for O&M has become increasingly severe. The problem has been exacerbated within the subsector by the continued development of new projects, whose capital cost is often funded from foreign grants or loans, and deteriorating elderly projects with component structures whose normal useful life has been exceeded. These twin pressures have increased the budgetary demands on national governments for O&M. Regrettably, rapid inflation of the last 10–20 years has dealt a final, almost fatal blow to the entire budgetary process. In these circumstances the budget 'game' deteriorates and becomes an activity with no winners. This can happen at the level of a national irrigation agency, whose budget must be approved and funded at higher levels of government, and at the level of individual projects within a national or regional irrigation agency.

Consider the plight of a conscientious project manager who devotes great effort to develop and submit a realistic budget, only to have the request arbitrarily cut in half by the funding authorities. He will certainly be in a worse position than another manager who submits a cursorily developed but well-padded budget, and who also receives a 50% cut.

We believe that a sound budget process is an essential prerequisite to good institutional management. One approach sometimes used to deal with this problem is to develop technical standards that permit the

BOX 4.1
Defining O&M standards for budgeting in Indonesia

In 1983 the Directorate of Irrigation of the Indonesian Directorate-General for Water Resources Development prepared guidelines for the budgeting of O&M expenditures. These guidelines were based both on a project's size and on the nature of its physical structures. Separate guidelines were prepared for each of the three major types of irrigation distinguished in Indonesia: technical, semi-technical, and simple.

For operations, the guidelines identify the number and type of personnel to be hired. For example, technical irrigation projects are divided into sections of approximately 30 000 ha. The guidelines call for each section to have six subsection offices with an office head and three staff. In addition, there is to be one irrigation inspector, one canal foreman and three people to guard weirs and gates for each 750 ha. The guidelines also specify the facilities that are to be provided for these personnel (office space, motorcycles, etc.).

For maintenance items, the guidelines identify both the typical density for various structures, and the maintenance cost of each structure. For example, a semi-technical irrigation project would be expected to have about one permanent weir per 750 ha, and 8 m of main canal per hectare. The annual maintenance cost to be budgeted for a permanent weir is *Rupiah* 120 million, and per metre of main canal is *Rupiah* 25 000.

Because the guidelines include the maintenance costs per unit for the various structures and facilities, it is possible for the irrigation department to calculate, on a more systematic basis, the amount of 'reasonable' O&M expenditures for the various projects. This is more realistic and systematic than simply budgeting a fixed amount per hectare of land irrigated.

calculation of a set of coefficients for 'proper' O&M (Box 4.1). International lending agencies and national irrigation agencies sometimes advocate or undertake this to encourage a more orderly and systematic approach to budgeting and funding of irrigation O&M.

This may be a fairly useful and convincing method of obtaining a more realistic assessment of O&M needs in situations where central financing exists; however, it carries with it the danger of the technological bias that we noted in the introduction to this chapter. Of course, even when top government planning agencies advocate the use of this approach, it still provides no guarantee that the funds will be forthcoming. The Indonesian case discussed in Box 4.1 is a good example. In spite of the use of technical standards for developing budget requests, the Indonesian government has not been able to fund fully the budget requests that have come to it from the provincial governments. In the early 1980s, for example, the government actually provided only about 60% of the funds requested.

There are many reasons for financial shortfalls from technical norms. The most prevalent is the failure of governments to obtain adequate revenue through the tax system, combined with an increasing number of demands on the limited funds available. These problems with the budgetary process and the increasing fiscal pressures that have faced national governments over recent years present powerful reasons to search for alternative ways to finance irrigation O&M. These alternatives generally involve some type of direct payment by water users for irrigation services and some degree of financial autonomy at the individual project level or at the level of a national irrigation agency.

4.2.2 *Funding O&M in the context of financial autonomy*

We shall argue that financial autonomy carries with it certain clear advantages for the financing of O&M. First, it frees, to an extent, those responsible for implementing O&M from the vagaries of government budgetary process, and potentially gives them access to greater resources. Second, particularly when an element of decentralised autonomy exists (e.g. at the level of the individual project, or of groupings of several small projects), the project managers are given powers and responsibilities that allow them to have considerable influence over the nature and quality of the irrigation service in their projects. This should substantially increase the professional rewards associated with irrigation O&M – a very important consideration. The widespread preference among engineers for construction over O&M may be traced back to elements from their initial training or attributed to the financial rewards in construction; however, it is reinforced by the poor pay, low status and inadequate resources associated with project operation. Finally, as is discussed more fully in the following section, financial autonomy tends to increase the influence of the water users in the O&M process.

But financial autonomy also carries with it potential problems and risks. If the irrigation agency is relatively unsuccessful in its efforts to assess and collect funds from the users, it could find itself with even fewer funds than might have been forthcoming from the government budget.

(The problems of collection and enforcement are taken up in more detail in Chapter 11.) This is likely to be a particularly difficult problem in the early years of a transition from central financing to financial autonomy, especially if the farmers are not accustomed to paying for irrigation services or if the quality of irrigation services during the previous years has been poor. It always has to be kept in mind that creating financial autonomy will require long-term political support, including careful planning and provision for a transition period during which some funds for irrigation O&M continue to flow to the irrigation agency from the government.

It is sometimes argued that because most farmers in developing countries are poor, either (1) they cannot afford to pay enough to provide for adequate O&M; or (2) it is politically unacceptable to ask them to pay very much towards O&M.

With respect to the first contention, irrigation farmers may be low income earners, but in relation to the rain-fed farmers, the landless, and the urban poor, they typically are privileged or at least less poor. A case for a subsidy to the less poor can be made; however, providing it through irrigation may not be the most appropriate means. In any event, despite their poverty irrigation farmers can generally contribute toward the public cost of providing the service they receive. Information on financially autonomous communal and private irrigation systems in various countries shows that even very poor farmers often pay quite large amounts for good quality irrigation services (see Chapter 10, Box 10.1). Indeed, in some circumstances they pay large amounts for poor services! If the quality of the irrigation services is very poor, it is of course likely that the farmers would benefit so little that they would be no better, or even worse off, if they paid for the O&M costs. But this is more a reflection of the failure to make the connection between payment and service quality than it is on the income level of the farmers. Even very rich farmers would resist paying an amount for irrigation services that exceeded the value they received from those services! The entire question of the farmers' ability to pay for irrigation services is examined in depth in Chapter 10.

The second contention noted above (of political acceptability) is best addressed by asking about the nature of the alternative to charging farmers for the O&M costs. Presumably it is seen to be politically unacceptable to charge the farmer because it is seen as imposing an economic hardship on poor people. This is a misplaced sense of responsibility. If the alternative is (as we shall argue is frequently the case) an irrigation system that operates poorly because of severe funding constraints at the national level, then 'helping' farmers by not charging for irrigation services can leave them worse off than if they were asked to pay for the services. If, in a given context, this is a valid description of the situation, and if this is understood by farmers and politicians, then their views regarding political acceptability of charges may change (Box 4.2). If the real reason for a lack of political will is that politicians see free or cheap water as a vote winner, the task is more difficult. We hope that the arguments of this text will strengthen the resolve and reasoning of political advisers who are fighting this misplaced charity.

4.3 *Assuring effective utilisation of resources provided for O&M*

'Don't bite the hand that feeds you.'
'He who pays the piper calls the tune.'
'Remember the Golden Rule: He who has the gold makes the rules.'

The above bits of folk wisdom attest to a central and obvious point of this book: financial arrangements have a profound effect on the behaviour of individuals involved in irrigation. The evident fact that is sometimes missed in irrigation policy discussions is that it is only the farmer who has the capacity to feed the system, to call the tune and to provide the gold. In this section we consider two facets of the question of how to assure effective use of resources provided for O&M. The first is that of accountability, and the second is that of expenditure decision-making.

4.3.1 *Accountability*

In general, the resources provided for O&M are entrusted to the agency responsible for operating the irrigation system. As a consequence, this agency is accountable for its use of the resources. Because accountability has many facets, however, it is useful to distinguish among three types of accountability.

(1) *Fiscal accountability* is a procedural concept concerned with the proper use of funds. Fiscal accountability is needed to ensure that the funds provided for O&M are spent in ways that are consistent with the purposes (usually as reflected in budget categories) for which the funds were made available to the agency, and in accordance with proper procedures. In terms of Fig. 4.1, fiscal accountability is designed to ensure procedural regularity in the processes whereby funds are used to acquire the physical resources for O&M.

(2) *Managerial accountability* refers to the judicious use of the resources provided to the irrigation agency in undertaking its responsibilities for O&M. It is thus concerned with the substantive aspects of both expenditure and resource-utilisation decisions (Fig. 4.1). With fiscal accountability the key question is

BOX 4.2
Applying economics to political arguments about irrigation charges: a hypothetical example

Economic analysts have a mode of thought that stresses consideration of alternatives. If water is free or very cheap it does not necessarily follow that the quality of service will decline. Government can allocate money from general revenue. On the other hand, if the money is provided it does not necessarily follow that O&M will be exemplary. However, we believe there is sufficient evidence presented later to show that the effectiveness of irrigation service is generally better when farmers pay a high proportion of total O&M costs to a financially autonomous irrigation agency.

A partial budget, such as in Table 4.1 below, can often be drawn up to show how it is rational for farmers to pay for irrigation, provided the money they contribute is spent wisely (see Alternative A) and not either wasted or lost in the central Treasury coffers (see Alternative B).

Table 4.1. *Farm yields and net income under alternative arrangements for financing O&M*

Item	Current situation: No charge to farmers for O&M, but low quality irrigation service due to low funding of O&M	Alternative situation A: Farmers pay for O&M but receive better quality irrigation service due to more funding of O&M and other benefits of financial autonomy	Alternative situation B: Farmers pay for O&M, but service does not improve
Crop yield	2.0 tonnes/ha	2.5 tonnes/ha	2.0 tonnes/ha
Net income before water charge	Rps 700/ha	Rps 1000/ha	Rps 700/ha
Water charge	0	Rps 100/ha	Rps 100/ha
Net farm income	Rps 700/ha	Rps 900/ha	Rps 600/ha

So why would the idea of farmer payments be politically unacceptable? Most likely, the situation as depicted in Alternative A in the table has not been recognised by the politicians, who are acting under the assumptions of Alternative B, i.e. that having farmers pay for the costs of O&M would simply transfer a cost now borne by the government to the farmers, with no change in the cost and quality of O&M, and thus no change in yields and income before payment of the water charge.

What is needed is an expanded perception of the changes that can be brought about by a shift to farmer responsibility for O&M in the context of financial autonomy. Water charges with good O&M could increase to Rps 300/ha and the farmers would be no worse off. Indeed, as we can expect technology to continue to shift the return to good irrigation upwards over time, we can expect the taxable capacity to increase.

---

whether proper procedures have been followed in making expenditures. With managerial accountability, on the other hand, the key question is whether an agency's decisions on expenditure and resource deployment (1) are appropriate to its purposes, and (2) achieve these purposes in an economic (non-wasteful) fashion.

(3) *Programme accountability* refers to an evaluation of the effectiveness of the actual O&M services provided. In contrast to fiscal and managerial accountability, which are concerned with procedures and decisions that are internal to the management process, programme accountability is concerned with the quality of the final output produced from the input of funds and managerial efforts. Programme accountability is thus the broadest of the three accountability concepts.

The relative emphases placed on each of these three types of accountability, and the direction and strength of the accountability linkages is likely to be affected by the way in which O&M is financed. In the case of central financing, the accountability linkages are primarily upward from the irrigation project to the central staff of the national irrigation agency and ultimately to the higher levels of government, such as a ministry of finance. Fiscal accountability linkages are likely to be much stronger than either those associated with managerial or programme accountability. This emphasis on fiscal accountability grows out of complex rules designed to prevent corruption and fraud in the use of public funds. One effect of these rules is, however, to make the fiscal process cumbersome, thereby slowing expenditures. In effect, the rules substitute for flexible management decision-making. It is much more difficult to devise rules to assure that expenditure decisions and utilisation decisions are appropriate and effective, and to assess the overall effectiveness of O&M. As a result, managerial and programme accountability linkages tend to be weak in situations of central financing.

The unfortunate result is that there is likely to be great concern within an irrigation agency to see that proper bookkeeping procedures are followed, but much less concern to see that the irrigation system is actually operated and maintained in a way that effectively meets the needs of the agricultural producers who use the water. Indeed, the needs of farmers and the ways in which they have changed over time are often unknown to the operators of the system. In some irrigation agencies the essential elements of the rule books are 100 years old and farmers are hardly mentioned.

This situation is likely to be radically different when irrigation agencies operate in the context of financial autonomy. The accountability linkages are now directed downward toward the water users who are required to pay, say, for the costs of O&M, rather than upward to the government. The water users are likely to be equally concerned with all three elements of accountability. Fiscal accountability remains important, because the farmers do not want corruption and misuse of funds to result in higher irrigation fees. But the farmers are equally concerned to see that the funds are judiciously spent in ways that they deem to be important for the operation and maintenance of the irrigation system, and that the resulting quality of O&M is satisfactory. By strengthening the managerial and programme accountability linkages, and by directing them toward the users rather than towards a government agency, financial autonomy creates incentives for the managers of the irrigation agency to give greater consideration to how their decisions and actions affect the quality of irrigation services provided to the users. Efforts to evaluate effectiveness of O&M, and to use those evaluations to modify resource utilisation decisions (see Fig. 4.1) are thereby strengthened.

4.3.2 *Expenditure decisions*

Under central financing, most expenditure decisions are likely to be independently made within the irrigation agency. Financial accountability linkages exist with higher levels of government, but at these levels there is little interest in or expertise on technical questions of irrigation operation. Furthermore, we argue that the lack of accountability linkages downward to the water users reduces the likelihood that farmers will be involved in these decisions.

With a degree of financial autonomy it is more likely that important decisions will involve both farmers and the personnel of the irrigation agency. Autonomy forces engineers and accountants to consult and obtain some agreements from the water users. If mutual recognition of needs is heightened, the irrigation agency is much more likely to be successful in collecting the funds that will be required for O&M.

An input of the water users' interests into the decision-making process is likely to influence a number of decisions on O&M. In particular, there are two broad areas where a proper consideration of the views of the water users is likely to lead to different results than would be expected when their views are excluded from the decision process. These are the questions of an appropriate maintenance strategy and of the appropriate O&M activities for farmers as opposed to hired staff.

Maintenance strategy

It is likely that engineers in an irrigation agency of the national government will tend to regard as appropriate a maintenance strategy that emphasises a high general standard of maintenance. Irrigation systems 'look nice' when they are maintained to such standards. The old military rule of 'if it moves oil it, if it doesn't paint it' is often applied. In some countries, where houses are never painted, trees and stones are whitewashed in almost ritual fashion by the irrigation department. But it is unlikely that such technical standards are economically desirable. Some parts of the irrigation system may change and even deteriorate considerably from the original design standards and still function quite satisfactorily.

In many agencies maintenance rules are inherited from long ago and accepted mechanically despite the changes that have occurred in engineering and technology. For example the advent of mechanised desilting and cleaning of canals, the adoption of short duration crop varieties, the increased availability of labour as a consequence of population growth, and the increased opportunity costs of recurrent budget finance are all factors that mean that technical manuals should be taken out from time to time, dusted and critically reviewed in the light of local endowment, and then rewritten. In our experience such a critical review is seldom undertaken. Where changes have been made to a project's crop production patterns, the need for reassessment of the operational norms is even more urgent. For example, in many systems the area irrigated has expanded as the value of water to farmers has increased, resulting in pressure on the operating engineers to run canals above their design level. This can raise operating costs by affecting things such as silting rates and increasing the risk of costly structural failure (e.g. canal breaches). This may well be an economic strategy, but it is unlikely to be in line with outdated operating manuals.

The economically optimal maintenance strategy should be responsive to field realities. It is likely to be one that permits selective deferred maintenance, i.e. one that allows certain parts of the system to deteriorate more than other parts prior to undertaking maintenance. Redesign and supplementary investments in more structures or in improved communication systems would be an important part of a responsive operation policy.

The determination of the details of a responsive operation strategy needs the input of both the farmer users of the irrigation system as well as the operators of the system. The farmers, who, under financial autonomy, must bear both the direct costs of the maintenance programme and its indirect costs (in terms of lower incomes) stemming from any deterioration in system performance caused by lack of maintenance, will have useful perspectives on the importance of various types of maintenance strategies. The more technically trained staff of the operating agency are in a better position to assess the risks and problems that may occur if maintenance of certain parts of the system is deferred.

Financing policies frequently create an additional complication for the selection of an appropriate strategy for maintenance. In many situations of financial autonomy, it is common for farmers to be responsible for the payment of recurrent costs (normally O&M), while the government pays for capital costs (initial construction and major improvements and rehabilitation). This creates an incentive to the farmers to accept deferment of certain types of maintenance to the point that the work can be considered to be rehabilitation of the system. Although this may cause the farmers to incur greater average costs because of poorer performance of the irrigation system, these costs may be more than offset by the private benefits that the farmers gain from reduced annual maintenance costs (Box 4.3).

Governments sometimes resort to a similar ploy when dealing with aid donors who will only finance the capital cost element of a project. Faced with this type of constraint on foreign aid, rational governments may first increase the capital cost proportion of total costs by demanding high specifications, and subsequently defer maintenance until deterioration

BOX 4.3
How financial policies can distort maintenance decisions

Expenditures are often categorised either as recurrent costs or as capital costs. When financial policies make water users responsible for the repayment of only the recurrent costs, an incentive is created for water users to find ways to shift costs from the recurrent cost category into the capital cost category. This can frequently be done by deferring maintenance of the irrigation facilities (a recurrent cost) to the point that a major rehabilitation of the facilities (a capital cost) is needed. Although such a strategy may create economic losses, it could still be in the best interests of the water users, as the example in Table 4.2 illustrates.

In this example, the water users suffer a decline in net income, prior to paying for maintenance, due to the poorer performance of the irrigation system resulting from the low level of annual maintenance. Furthermore, the total cost of maintenance (including rehabilitation) is greater under the deferred maintenance strategy. From the perspective of society, the deferred maintenance strategy is inefficient, as it generates an average annual net income that is £20 below the annual maintenance strategy. But because the deferred maintenance strategy allows the users to shift an annual cost of £35 to the government, the users can still gain by an amount of £15 by following the deferred maintenance strategy.

Table 4.2. *Maintenance strategy*

	Annual maintenance	Deferred maintenance
Annualised present value of water users' net income per ha, prior to paying maintenance costs	£100	£90
Annual maintenance costs[a]	£30	£5
Annualised present value of rehabilitation costs[b]	£0	£35
Annualised present value of net income, including all maintenance costs	£70	£50
Annualised present value of net income received by water users	£70	£85

[a] Paid for by water users.
[b] Paid for by government.

justifies rehabilitation – which is a form of project assistance that qualifies for foreign aid as a capital cost.

Roles of the water users

Responsibility for undertaking the O&M activities could be given to the water users, could remain with the irrigation agency, or could be shared. In the first case, each individual farmer would have the option of either doing the work with family labour, or hiring someone else to do it. If the irrigation agency were fully responsible, it would hire labour in the form of either regular staff, daily paid staff or contract labour to undertake the work.

In situations of central financing, where the water users are not required to pay for the costs of O&M, it is obvious that they are likely to expect the irrigation agency to be responsible for these activities. Irrigation agencies, faced with inadequate budgets to undertake all the desirable O&M activities, often plan to rely upon water users for the maintenance of small ditches, watercourses and field distribution channels. Conflicts between the agency and the water users frequently develop over the question of how much work should be diverted from the agency to the farmers. Throughout the irrigation world we find that maintenance work does not get done adequately in situations where responsibilities are diffuse.

If financial autonomy prevails, and water users pay at least a substantial proportion of the O&M costs, the situation is likely to be quite different. In the textbook situation of perfect markets and no transaction costs, we could expect the water users to be indifferent as to whether they or the irrigation agency had responsibility for the work. They would either pay cash for the work or else undertake it themselves.

In reality, imperfections in the market and the existence of transaction costs may make the water user prefer one approach over the other. Imperfections in the labour market are likely to mean that the opportunity cost to the farmer of his or her labour (or of the labour of some other family member) is less than the rate at which the irrigation agency would have to hire labour. As a result, the water users may prefer to undertake a number of tasks themselves, rather than rely on the irrigation agency for these services.

Transaction costs associated with the supervision of labour may also give a cost advantage to having certain types of work done by the water users. Workers hired by an irrigation agency to undertake maintenance on small channels need supervision. If the agency hires supervisory personnel, explicit cash costs are incurred that must be passed on to the water users. Alternatively, if the agency fails to hire supervisors, the quality of the work may suffer, which means that the water users are paying for inferior work. It is reasonable to expect that water users can undertake many local O&M activities with lower supervision costs. In part this reflects a decreased need for supervision because the work is being undertaken by those who have a direct interest in the quality of the work. To the extent that supervision is needed, it can probably be done more informally (because the farmers already know each other) and therefore at a cost lower than that of the irrigation agency.

Irrigation agencies frequently have to pay statutory minimum wages, set to satisfy urban consumption needs, that are much higher than informal rural wage rates and thus higher than society's labour supply price, which reflects the opportunity cost of labour. Hence labourers hired at minimum wages enjoy economic rent and this explains the long queue of labour at construction camps.

It would be incorrect to assume, however, that the water users are in a position to undertake all the O&M activities. They are likely to lack the experience and expertise needed to undertake O&M activities related to some of the more complex structures or equipment of the project. Furthermore, despite seasonal unemployment at certain times of the cropping calendar, the opportunity cost of labour on farms is very high during critical periods of peak activity. The challenge is to try to find, for any given irrigation system at any particular point in time, the types and timing of tasks that are best suited for the water users to execute, and those that are best suited for the staff of the irrigation agency. This requires the type of interaction between the water users and the personnel of the irrigation agency that is seldom achieved, but which can be encouraged by financial autonomy. This is a radical step. Once farmers have been asked to share the burdens of cost, they are likely to seek to participate in management decision-making. We believe irrigation staff should actively encourage this.

4.4 *Summary*

In assessing the present poor state of irrigation management and suboptimal performance there is a great temptation to use spurious reasoning and to adopt the 'fallacy of the opposite'. Poor revenue collection is associated with poor operation and maintenance and operational performance; therefore, it is often assumed that good revenue collection will lead to good operations. Additional revenue may or may not be necessary for good operation, but it certainly is not a sufficient condition. In fact, it is more likely that good operation and maintenance will lead to good revenue.

Effective operation and maintenance of the physical facilities is also necessary if irrigation investments are to realise their potential. Two key but interrelated elements of financial policy must be addressed: how to obtain the funds needed to operate and maintain the facilities; and how to ensure the funds obtained are actually used to provide an effective and efficient irrigation service.

Centrally financed irrigation agencies are often subject to arbitrary budget cuts reflecting the fiscal problems of the central government. Financially autonomous agencies avoid this difficulty, but are faced with the major challenge of establishing an effective system of user fees whereby small amounts of money are obtained from large numbers of small and often low-income farmers.

A system that is to ensure effective and efficient irrigation must establish proper lines of accountability. One of the major advantages of financial autonomy is the potential that is created through the system of user fees for the accountability linkages between the managers of the irrigation agency and the water users. These linkages may help give water users a voice in determining how the agency's funds are to be used, thereby involving them in the process by which the total 'need' for funds is decided.

5

Allocating a scarce resource: water-use efficiency

How often one hears the complaint that irrigation water is wasted! Farmers who are favourably located along the irrigation canals tend to use more water than 'needed', frequently increasing the problems of shortages and unreliable supplies for the tail-end farmers. The resulting 'inefficient' allocation and use of water seems obvious to many irrigation professionals. Engineers note that technical water-use efficiency (the ratio of the amount of water actually used by the plants to the amount of water entering the irrigation system) is very low. Irrigation agronomists focus on the large differences between the amounts of water applied to farmers' fields and the amounts needed by the crops for optimal crop growth.

Economists, however, have a different (and in some ways more difficult) approach to evaluating inefficiency in irrigation. They evaluate efficiency in the allocation and use of water from the broader perspective of economic efficiency. Water must therefore be considered, not as an isolated and separable input used in agricultural production, but as one of many interrelated inputs whose use must be optimally integrated in a flexible production 'package'. Prices (be they market prices or shadow prices) become the common denominator by which the various components of this production package can be compared, and through which efficiency can be evaluated.

Many economists, noting both the 'waste' of irrigation water and the frequency of large subsidies given in ways that make the use of this water free at the margin, have concluded that much of the waste is due to inappropriate water prices. Farmers with good access to irrigation water use 'too much' because they are responding in a rational economic fashion to its low or zero price. At this price, it is economic for them to substitute water for other higher priced inputs such as labour. This suggests that a major cause of economic efficiency in irrigation is that the marginal cost faced by the individual water user does not reflect society's true opportunity costs.

In this chapter we explore the prospects and limitations of using irrigation financing policies to cause farmers to behave in ways that will foster economic efficiency in the use of water. We begin in the following section by noting the importance of prices for creating the desired incentives to economise on water use, and by considering key factors that affect the extent to which systems of water prices can lead to these desired results. In the subsequent section, we turn to an examination of some of the requirements for establishing water pricing mechanisms. Finally, we discuss the conceptual question of the appropriate level and structure of prices for maximising economic efficiency.

5.1 *Prospects for water charges to increase efficiency*

5.1.1 *Effects on water-use decisions*

A necessary condition for an irrigation financing mechanism to improve the efficiency of water use is that it have some influence on farmer decisions affecting water use. Farmers make three basic types of such decisions. *Cropping decisions* involve the number, type and timing of crops grown. They affect water use by modifying the amount and timing of the demand for water associated with agronomic characteristics of the crop. *Water conservation decisions* affect the conservation of water on the farm, thereby affecting the technical efficiency with which irrigation meets these demands of the crop for water. *Water acquisition decisions* are the ultimate decisions about the amount of water to acquire for the farm. They reflect the farmer's perceptions about the optimal amount of water stress under which to grow the crop, but are also affected by both the cropping decisions and the water conservation decisions. Relationships among these decisions and the factors affecting them are indicated in Fig. 5.1.

Cropping decisions are likely to be influenced both by factors that directly affect the expected profitability of alternative potential crops and by other socioeconomic factors such as household consumption preferences, reliability of markets, and off-farm employment opportunities (Fig. 5.1). Area-based fees that are differentiated according to the crop grown may affect these decisions through their effects on the expected

Fig. 5.1. Relationships between the cost of water and farmer decisions affecting water use. Factors in the circles are external factors outside the control of the farmers. The three items in the diamond-shaped figures are the three types of water-use decisions that farmers make. The items in the rectangles represent actual outcomes resulting from the interactions of the farmers' decisions and the external factors.

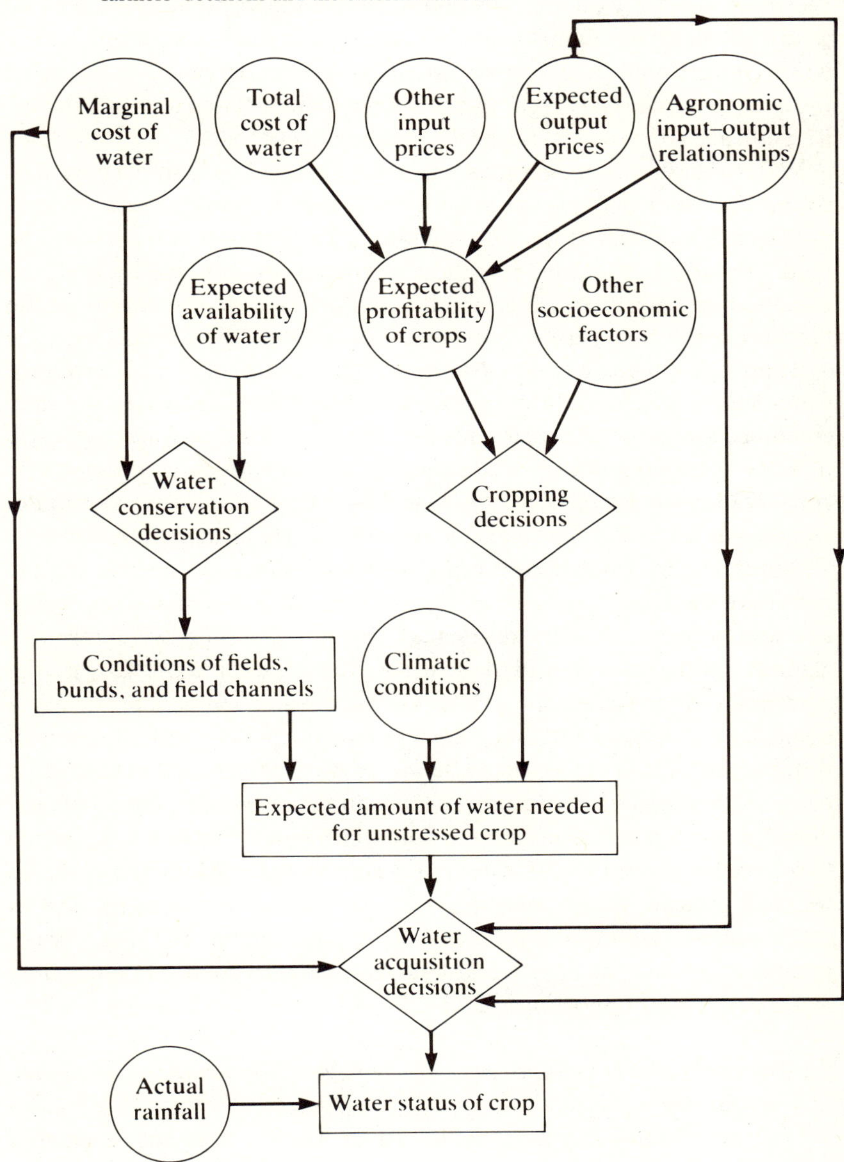

relative profitability of the crops. However, as was noted in Chapter 3, in many cases the cropwise differential in the charge for water is too small, relative to other factors affecting profitability, to have much effect on cropping decisions (see Box 3.1). A few crops, however, with rice being the most notable example, are likely to use much greater amounts of water, particularly under some soil conditions, than most crops. If the water charges realistically reflect this difference, and if the overall level of the charges is relatively high, then a system of area-based fees might have a significant effect on the farmers' cropping decisions.

Water conservation decisions are largely related to land preparation, the maintenance of field channels for water distribution on the farm, and, in the case of irrigated rice, the maintenance of bunds surrounding the field. As indicated in Fig. 5.1, these decisions are affected by both the expected availability of water and the marginal cost to the farmer. If the farmer expects water to be very scarce, or if its marginal cost is high, he will find it profitable to incur costs, largely in the form of labour, to reduce water losses by carefully preparing the land and maintaining the field channels. Only if the irrigation fee is in the form of a true water price will it have the potential to reduce water use by affecting these water conservation decisions. An area-based fee, regardless of its magnitude, will always leave the marginal cost of water to the farmers at zero.

Water acquisition decisions are made at a number of times during the growth of the crop. Application of irrigation water is part of a sequential decision process. At any particular time in this process, a farmer's acquisition decision would depend on the type of crop grown, the marginal cost of water, the extent to which water stress is expected to result in a reduction in yield, the relationship between the amount of water acquired and the expected degree of water stress, and the expected price of the crop. As in the case of the water conservation decisions, the only financing mechanism with the potential to cause the farmer to decide to reduce the acquisition of water is a system of water prices. However, as we shall indicate in the following section, if water is scarce relative to other inputs, the farmer will face an implicit opportunity cost when applying the water to a particular field, even in the absence of an explicit water price. This opportunity cost will guide rationing decisions.

5.1.2 *Administrative rationing and the value of irrigation water to the farmer*

Irrigation water has a value to a farmer that does not necessarily bear any relationship to the amount that he pays for it. This value depends on three factors: the technical efficiency with which the irrigation water delivered to the farmer is made available to the crop; the biological response of the crop to water; and the price of the crop. Taken together, and considered at the margin, these three factors define what economists call the value of the marginal product of water, i.e. the monetary value of the additional output generated by one additional unit of water.

From economic theory we know that the profit-maximising amount of input use occurs where the value of the marginal product of the input equals its marginal cost. This simply means that as long as there is no constraint on the availability of water, a farmer will find it profitable to continue to increase the amount of water used until the resulting additional revenues drop to the point where they exactly equal additional monetary costs.

But why will the value of the marginal product decline with an increase in the amount of water applied? The answer lies in the biological response of the crop to water. When the crop is suffering from severe water shortage, the crop will show a highly positive response. But as additional water is supplied, the increase in production from a given increase in water will be smaller. Eventually if enough water is added, a point will be reached where the total amount is actually in excess of the crop's biological optimum. At this point the crop's response to the additional water will be zero or even negative (see Box 2.1). This changing biological response is, of course, nothing more than another example of the ubiquitous rule of diminishing returns.

So, to repeat: in the absence of a constraint on the availability of water, a farmer will find it profitable to use water until the resulting additional revenues drop to the point where they exactly equal his additional monetary costs.

Thus far, our discussion in the above paragraphs is nothing more than an elaboration of the economic logic underlying our previously stated conclusion that an increase in the price of water would cause the farmer to reduce the amount of water he or she chooses to acquire. Furthermore, the discussion points out the fact that under the assumed conditions, each farmer would tend to behave in such a way that the 'value' of water (or, to be more precise, the value of its marginal product) would be equal to the price of the water.

But what if the supply of water is restricted? This is certainly the case in many irrigation projects. For example, irrigation projects in parts of India and Pakistan operating under the Warabandi system of water distribution were designed to spread a limited supply of water over an area roughly three times as large as the water supply could irrigate. To the extent that the distribution system functions as intended, each farmer is thereby restricted to a total water supply that is too small to irrigate the entire farm. Now let us assume further, as is also generally true for irrigation systems in India and Pakistan, that farmers pay area-based fees for water, rather than water prices. The farmer's marginal cost of acquiring water is thus zero.

Under these conditions, each farmer must decide how large an area to plant and irrigate with the limited supply of water. Although one choice might be to plant such a small area that the value of the marginal product of water would be zero, it would be foolish to do so! The reason is simple: although the marginal cost of acquiring the water from the government is zero, the scarce water has a very high marginal opportunity cost to the farmer. When part of the farmer's land must remain fallow for lack of water, the net value of the crop that could have been produced on that part of the land becomes the opportunity cost of the water. A profit-maximising farmer can be expected to attempt to allocate limited water supply in such a way that the crop produced will suffer an 'optimal' amount of yield reduction due to water stress. In some circumstances (lots of land, cheap labour and low input levels), it will be optimal to spread the water very thinly, resulting in relatively low yields over a relatively large area. By allocating water in this fashion, the profit-maximising farmer will cause the value of its marginal product to be approximately equal to the opportunity cost of the water at the farm level.[1]

As long as administrative rationing results in an opportunity cost of water to the individual farmer that is greater than the actual or proposed water price, a price system cannot be expected to increase the efficiency with which water is used. In their role in promoting economic efficiency, prices act as powerful signals of opportunity costs, and thereby become the mechanism for rationing. But if water is effectively rationed administratively so that its opportunity cost is higher than its price, prices lose their rationing function, and can no longer be expected to be an effective means for enhancing economic efficiency.

5.1.3 *The value of irrigation water to society*

The extent to which water prices may increase the overall economic efficiency of water use also depends on the value, or opportunity cost, of the water to society. This opportunity cost may reflect foregone opportunities for additional irrigated agricultural production, or it could reflect foregone opportunities to use water in other sectors of the economy, such as for industrial production or household consumption.

When water is relatively abundant, its opportunity cost approaches zero. This may occur in run-of-the-river diversion systems where the total irrigable area is small relative to the volume of water in the river. It may also occur during wet seasons in the semi-humid tropics. In these situations any price for water in excess of society's marginal cost of distributing it would reduce economic efficiency by discouraging farmers from using a resource that cannot be saved and that has no alternative value.

But when water is scarce, the situation is totally different. In an economic sense, scarcity implies that water has a high opportunity cost. Scarcity thus depends not simply on the amount of water available, but also on the nature of the demands for it. Industrialisation and urbanisation will increase the economic scarcity of water, as will agricultural investments in things such as improved varieties, crop protection or land levelling. These investments shift the water response function upward, thereby increasing the average and often the marginal return to water. Under conditions of scarcity, it is the reality of water's high opportunity cost that needs to be conveyed to potential water users. A system of water prices reflecting these opportunity costs may therefore enhance economic efficiency.

These differences in the value of water to society help explain why systems of water prices are much more commonly found in the water-scarce arid and semi-arid regions of the world than in humid and semi-humid regions.

5.1.4 *The elasticity of demand for water*

The efficiency gain from a system of water prices depends on the degree of responsiveness of water users to the prices charged. If there is little change in water use in response to a change in the price of water, then a system of water prices will generate little gain in efficiency. The responsiveness of the water users to the price of water is measured by the elasticity of demand for water.

Many irrigation experts (usually non-economists) tend to feel that there is little or no price responsiveness for irrigation water. This thinking reflects a tendency to focus on concepts such as 'water duty', 'irrigation requirements' and 'crop-water requirements'. But economists are much more likely to expect that, as with other inputs, farmers will show some degree of price responsiveness for irrigation water.

Little empirical information is available on responsiveness to water prices in the developing world. In part this is because systems of water pricing in these countries are rare. Water prices are most likely to be found in private and public pump projects, and it would be useful to have studies on the price elasticity of demand for water in these types of projects. It must be recognised, however, that the generalisations that can be drawn from such studies may be limited because the water supply in these kinds of systems is generally more reliable than the supply in large gravity systems. As a result, the farmers' responses to water prices may be rather different.

Where studies have been done, however, they tend to show that water users respond modestly to water prices. Studies in the semi-arid western United States, for example, have found the demand for water to be price inelastic, with estimates of elasticity of demand generally between −0.6 and −0.7. This suggests that water prices could have a significant effect on the efficiency of water use, with a 10% increase in the price of water reducing demand by an estimated 6–7%.

5.1.5 *The cost of implementing a price system*

Even if a system of water prices were to increase the efficiency of water use, overall economic efficiency would not necessarily be enhanced. The gains associated with the increased productivity of the water must be compared with the costs of implementing the system of water prices. The price system will enhance overall economic efficiency only if the gross economic gains are great enough to give a return on the cost of implementation equivalent to the rate of return that could be expected from alternative investments of government funds.

The gross economic gains attributable to the price system consist of two components. The most obvious of these is the increase in value-added from the irrigation water because of its reallocation. The second component, which is easily overlooked and extremely difficult to quantify, is related to the economic distortions that are induced when taxes are imposed to finance irrigation. To the extent that a system of irrigation pricing allows a reduction in the use of taxes to finance irrigation, these distortions will be reduced. The question of the marginal social costs of taxation associated with the financing of irrigation subsidies is a greatly under-researched area. (See Chapter 7 for further discussion of this matter.)

In most gravity irrigation systems serving large numbers of small farmers, implementation costs for water pricing systems are likely to be very large. These costs of assessment, billing, collection and enforcement are dealt with in greater detail in Chapter 7. It is worth noting here, however, that if the assessments are based on actual volumetric measurements of water deliveries to individual farmers, the implementation costs could easily be so high that they outweigh the efficiency gains of the price system.

5.2 *Key requirements for water pricing mechanisms*

5.2.1 *Ability to make selective water deliveries*

Any system of prices implies the ability of the seller to prevent delivery of the good or service to those who are unwilling or unable to pay for it. In the case of irrigation, the operating agency must thus have the ability to deliver water to those units (individual farmers or water users' associations representing groups of farmers acting together) who are willing to pay for it, while simultaneously excluding all other units from access to it.

In addition to the emphasis on exclusion, a system of irrigation pricing also necessitates that the irrigation agency be reasonably responsive to the desires of the farmers with respect to the timing and amounts of water delivered.

A price system thus places strong pressures on an irrigation agency to deliver water according to patterns that to a significant extent are determined by the water users rather than by the agency. Most irrigation systems serving large numbers of small farmers, however, have limited ability to deliver water selectively to some farmers while denying it to others. In such situations, efforts to establish an effective system of water prices are unlikely to succeed.

5.2.2 *Measurement*

Cloth is sold by the metre. When we buy a piece of cloth, its length must be measured to determine how much we must pay. If the price of rice is 10 Pesos per kilogram, the rice that we purchase must be weighed. If the price is given in terms of Pesos per litre, then the volume of rice that we purchase must be measured. Any system of price implies the need for measurement.

Measurement of water in volumetric terms would permit the establishment of a system of water prices that could be expressed in monetary units per cubic metre of water. But measuring irrigation water flowing in open channels presents a number of difficulties. Although techniques to measure rates of flow per unit of time exist, obtaining accurate measurements under field conditions, particularly where rates of flow are small, may be both difficult and expensive. Farmers may not cooperate in maintaining measuring devices. They might even deliberately destroy them! Furthermore, flow rates can be converted to measures of the volume of water only if information is available on the length of time that water flowed. If the rate of flow varies over time, it may be necessary to measure it frequently in order to determine accurately the total volume of water delivered.

If we can ignore implementation costs for the moment, volumetric measurement of water represents the ideal approach to pricing irrigation water. There appears to be a technology gap at present, however, and so pricing enthusiasts must await a low-cost tamper-proof device that can accurately measure silt-laden water in harsh field conditions. (Such a challenge deserves to be attacked by research teams, for the potential market for such a device with, say, one per 20 hectares of irrigation is more than 10 million!)

Fortunately, volumetric measurement is not necessarily required for an effective water pricing system. The key element in a system of water pricing is the farmer's ability to affect the amount to be paid for water through decisions on water use. For this to occur, the amount of water received by a water user and the costs for water must vary in a reasonably predictable fashion with water-use decisions. It is not necessary, however, that water payments be strictly proportional to water deliveries. Proxies for the volume of water may be used. Possible proxies include the length of time of delivery, the number of times a crop is irrigated, and the share of a variable water supply to which a farmer is entitled.

For example, in some pump irrigation projects, a water user is charged according to the length of time that the water is received. If the price per hour were constant throughout the cropping season, and if the volume of water delivered per hour were to decrease over the season (due to, say, a declining water table), then the effective price per unit of water would rise over the season. It is unnecessary to measure the actual volume of water for this pricing system to encourage the water user to be efficient in his use of water. For such a system of time-based pricing to work well, however, it probably is necessary that there be some reasonable day-to-day consistency in the volume of flow delivered per unit of time. Otherwise, a system of pricing by time would likely be viewed as unfair and become unworkable. This consistency in volume of flow is, unfortunately, lacking in many gravity irrigation systems.

Another proxy for volume that might be used as the basis for a water pricing system is a share of the flow. Although the total flow might not be known, either in advance or after the fact, an individual would be entitled to, and pay for, a certain share of whatever flow exists. Traditional communal irrigation systems sometimes have allocated water rights to their members on this basis (Box 5.1). Another proxy for volume occurs when irrigation water is made available for discrete irrigations of a more or less known amount. A farmer can thus decide whether or not to purchase an additional 'watering' even though the precise volume of water delivered is not measured.

Part of the difficulty with volumetric pricing is the large number of measurements that must be made when the billing units consist of very small farmers. This suggests the possibility of volumetric pricing of water on a 'wholesale' basis, whereby water is delivered and sold 'in bulk' to organised groups of farmers at points in the system where the measurement of volume is feasible. The farmers' organisation purchasing the water would then be responsible both for its ultimate distribution to the individual water users, and for the collection of water charges from them.

This type of 'water wholesaling' would obtain some of the benefits of volumetric water pricing without incurring unreasonable physical, administrative and financial burdens. It might also encourage more efficient operation of the irrigation facilities by placing pressure on the irrigation agency to make deliveries at the specified points in accordance with agreements made with the water users' organisations. Its primary weakness lies in the fact that unless the number of farmers in the purchasing group is reasonably small and cohesive, the individual water users may continue to behave as if the marginal cost of water were zero. There is some inconclusive research work on the optimum size of an irrigation user-group but this is surely a fertile field for sociologists and anthropologists to explore locally appropriate solutions.

Having noted that actual volumetric measurement is not a necessary condition for a system of water prices, we must now point out that volumetric pricing is also not a sufficient condition for creating water prices. In other words, a system of volumetric charges for water does not guarantee that the requirements of water pricing have been met. If the water user has no control over the volume and timing of the water received, then charging for water volumetrically would have no influence on his or her water-use decisions. For example, the Warabandi system of

---

BOX 5.1

Water pricing using flow shares: an example of a communal irrigation system in Nepal

Edward Martin and Robert Yoder undertook an intensive study of two communal irrigation systems in Nepal. The Chherlung Thulo Kulo (CTK) system was developed on the basis of issuing shares to farmers in proportion to their contribution to the initial investment costs of the system. A total of 50 shares were originally issued to the 27 households that contributed to the initial investment, with each share entitling the holder to 1/50 of the total water delivered by the system. The result of the initial distribution of the shares was that some farmers were entitled to more water than they needed, while others wished to increase their access to water. Still other farmers who had not contributed at all to the investment in the system, but whose land could be served by it, decided they would like access to the water. This led to market transfers of shares and portions of shares. The price of these shares has risen significantly over the years.

Over time, improvements to the irrigation system increased the total amount of water delivered by the system. Many farmers holding the original shares found it more profitable to sell some of their shares and be more conserving of water than to retain them and enjoy the enhanced water supply. The area irrigated has expanded, and the overall technical efficiency of water use in the system has become very high. Quite clearly, the allocation of water on the basis of shares that have a high market value has created a high opportunity cost for water to the farmers in the CTK system. This in turn has given them incentives to use water carefully so that they can reduce the number of shares that they need to hold.

The second system studied by Martin and Yoder, the Argali Raj Kulo (ARK), operates on the basis of water rights that are tied to specific parcels of land. Like the CTK system, the total water supply to the ARK system has increased over time. But with no potential for selling water rights, the improvements have led to less expansion of the irrigated area than occurred in CTK. Instead, these improvements have allowed those farming the irrigated land to enjoy the benefits of more abundant water.

Based on data that they collected over an entire irrigation season, Martin and Yoder estimate that relative to the demands of the crop, water use per hectare in ARK is about 30% greater than in CTK. They conclude that the system of marketable water shares has led to the more efficient utilisation of water in CTK.

---

water distribution used in north-western India and Pakistan has a rigid pattern of timed turns for water delivery to individual farmers. Although payment for water in these systems generally involves area-based fees, the suggestion has been made that assessments be converted to an approximate volumetric basis. But doing so would entail primarily an accounting change in the billing procedures for the water fee, and would not create a true system of water prices. As under the current system of area-based fees, farmers would still not be able to affect either the amount of water received or the amount paid. For this reason, a change to this proposed system of 'volumetric' water charges would not enhance the efficiency of water use.

5.3 *'Fine tuning' a water price system to promote efficiency*

Although a system of water prices may increase the efficiency with which water is used, economic efficiency will be maximised only if the prices are set at 'the right' level. But what, at least in concept, is 'the right' price for water? And should a single price prevail within an irrigation project or from an individual tubewell, or should a price structure be established within which the actual price can vary over time and space?

5.3.1 *What is the 'right' price?*

Many economists who have addressed the question of the 'right' price (in terms of economic efficiency) for water conclude that as long as excess capacity exists, the price of irrigation water should equal the marginal cost of providing it, but whenever a capacity constraint exists the price should be allowed to rise above marginal cost to the point where the quantity demanded just equals the available supply.

Underlying this conclusion are two concepts: (1) that maximising the net benefit of irrigation water to the economy requires that the marginal benefits to all users be the same, and (2) that in the absence of a capacity constraint, the marginal benefits of the water to the economy should equal the marginal cost to the economy of providing the water. Rationing the supply by means of a common price to which all users attempt to equate their marginal benefits addresses the conditions underlying the first concept. Administratively setting this price at the marginal cost of supplying the water meets the second criterion as long as this completely rations the supply available. But if a capacity constraint exists (which means that when the price is set to equal the marginal cost, the quantity of water demanded exceeds the supply available) the only way to meet the first criterion is to raise the price to the level that will exactly equate total demand with total supply.

The above conclusion concerning the right price for water is essentially a conclusion in favour of 'marginal cost pricing', a term used by economists to mean setting the price of a product equal to the incremental costs associated with increasing its production by an additional unit.[2] But difficulties associated with marginal cost pricing have led some economists to question the appropriateness of this concept for pricing irrigation water.

The general use of marginal cost pricing throughout the economy would require large government subsidies in many sectors. This is because marginal costs are often below average costs – a situation that occurs when the average costs of a firm (or of an irrigation agency) decline as the size of the firm increases. This would probably be the case in much of the distribution sector of the economy. Full adherence to marginal cost pricing would thus require a considerable expansion in the scope of activity of the government.

Financing these subsidies would create inefficiencies in the economy. Advocates of marginal cost pricing usually argue that the subsidies would best be financed by income taxes, since other types of taxes (such as excise taxes) would tend to create distortions in the relative prices of goods. But in its effects, an income tax is a form of excise tax on work. By making work less remunerative, and thereby reducing the opportunity cost of leisure, the tax distorts people's allocation of time away from work and toward leisure. It is therefore not certain that a system of marginal cost pricing in the economy, financed by large income taxes, would be better than a pricing system based on average costs with no need for subsidies.

For some types of goods (including irrigation water) one possible solution to the financial problems created by marginal cost pricing would be to establish a two-part charge for the product. One portion of the charge would be based on marginal cost, and would be levied on each unit of the output that is purchased. The second portion would be a lump sum charge levied in such a way as to avoid distortions of individual decisions regarding the use of the product. By setting this charge high enough to cover the difference between total costs and the revenues generated by the sale of the product at its marginal cost, the need for a subsidy would be eliminated.

Although the incorporation of marginal cost pricing in a two-part charge for irrigation water could presumably solve the subsidy problem, economists can still not be certain that this is the most desirable approach. The problem is that many prices in the private sector of the economy are not set at the marginal costs of production. Given this fact, should the government try to achieve economic efficiency in the allocation of irrigation water by setting its price equal to its marginal cost? The idea seems attractive on the surface; however, the 'Law of the Second Best' weakens the presumption that doing so will increase efficiency.

The Law of the Second Best says that when several prices deviate from what is optimum (the 'first best' condition), changing one of them toward the level that would prevail under the first best or optimum situation may not be an improvement. For example, the 'second best' solution in a given situation might involve several prices all deviating from their optimum amount by the same percentage. Starting from this second best (non-optimal) situation, changing any one price to what it should be in the optimum condition would actually make things worse, rather than better.

Another difficulty is that when one begins to examine in some detail the concept of the marginal cost of irrigation water, it becomes clear that there are a variety of marginal costs. The question that then arises is which of these marginal costs should be used in marginal cost pricing? The fundamental answer in conceptual terms is that the price should make the consumer's incremental cost (paid as a result of a consumption decision) equal to the additional cost that this decision requires the supplier of the product to incur.

The question of the appropriate price depends therefore on the nature of the irrigation decision with which we are concerned. In the case of the development and utilisation of an irrigation system, many different kinds of decisions can be identified, each of which would involve its own marginal cost. Examples of operational decisions include: (1) turning on an entire main irrigation system; (2) turning on an individual distributary unit within an irrigation system; (3) providing more water to an individual distributary unit; (4) providing more water to a specific individual in a distributary unit; and (5) reallocating water deliveries so that one part of an irrigation system receives more water while another part receives less. There are also several types of investment decisions that might be made, including: (1) adding a distributary unit or a group of such units to an irrigation system; (2) adding capacity to the main irrigation system; and (3) improving the channels in a distributary unit.

The marginal costs to which the public agency running an irrigation system should be responsive would be different for each of the above decisions. Furthermore, in many cases the opportunity cost of water comprises a large component of the marginal cost. For example, delivering more water to one part of an irrigation system (Part A) while reducing delivery to another section (Part B) means that the additional water delivered to Part A carries an opportunity cost equal to the net value of the reduced production in Part B. These opportunity costs will also depend on the type of decision being made.

We feel that the question of the 'right' price for irrigation water in developing nations needs to be put in perspective. In reality, true water pricing is seldom used. In those situations where pricing is possible, the establishment of the price system itself – giving water a positive marginal cost to the users – is probably far more important than is the precise level at which the price is set. Furthermore, if financial autonomy is a real possibility, then it would be important to design the pricing system to generate enough revenues to cover the irrigation agency's recurrent O&M costs.

For those situations where (1) financial autonomy prevails; (2) implementation of true water pricing is feasible at a reasonable cost; and (3) significant efficiency gains can reasonably be expected from water pricing, we would favour a two-part charge: a water price based on volume or some proxy for it, combined with a fixed charge based on the area irrigated. We favour this approach for two reasons. First, it would give greater year-to-year stability to the revenues of the irrigation agency than would a pricing system that did not include a fixed charge. Second, it would allow the irrigation agency greater flexibility in setting the actual water price at whatever level seemed most appropriate in a given situation. In many cases this would be considerably below the level that would be necessary if the price were the only component in the irrigation charge. This would be true in situations where water is fairly abundant and the irrigation agency's marginal operating costs are relatively low. In other cases, particularly where water is scarce and its opportunity cost high, the price could very well be higher than would be necessary to generate the funds to cover the agency's operating costs. In such cases the financial arrangements could call for a portion of the funds collected to be used to pay for part of the investment costs.

5.3.2 *Should the price of water vary over time and space?*

As we have previously stated, basic economic concepts indicate that maximising irrigation's net benefit to the economy requires equating the marginal benefits of all users, and that this can be achieved if all users face a common price. This assumes, however, that the marginal cost of supplying water to all users is the same. In fact, the marginal cost of irrigation water generally varies over both space and time. Seasonal differences in the availability and in the demand for water may create differences in marginal costs. When water is scarce and the demand is high, special efforts may be necessary to effect the desired deliveries. Furthermore, these conditions imply a high opportunity cost of the water itself. But the same irrigation project may experience periods of time when water is relatively abundant, making deliveries easy and giving the water a low opportunity cost.

If marginal costs vary over time, then the underlying economic logic of a pricing system suggests that, other things being equal, it would be desirable for the price of water to vary over time. One possibility would be to use a peak period pricing arrangement. For example, in rice-based irrigation systems, the charge for water during the peak demand period of land preparation might be higher than during the rest of the season. This would create incentives for some farmers to postpone land preparation in order to obtain cheaper water. Careful evaluation of such an approach would be necessary, however, because this 'solution' might create even greater problems because of the resulting lack of uniformity of crop growth in the irrigated area. Another example relates to daily fluctuations in demand. Most farmers prefer to irrigate in the daytime, creating a greater demand for water than exists at night. This suggests that a lower price for irrigation delivered at night might enhance efficiency.

Marginal costs also differ over space. Because of seepage and percolation losses in the irrigation channels, the real resource cost to society of delivering one unit of water to a farmer at the tail end of an irrigation system is greater than the cost of delivery to a farmer near the source of the water supply. If, for example, half of the water diverted at the head of the system is lost while in transit to the tail reaches, then the marginal opportunity cost of one unit of water received by the tail-end farmer would be approximately twice that of one unit of water received by a head-end farmer. If marginal cost pricing were to be followed strictly, volumetric prices charged to farmers on the basis of actual deliveries would thus have to vary within a single irrigation system according to location.

As previously noted, if other things are equal these suggested spatial and temporal price differentials would increase economic efficiency. But clearly not all other things are equal. In particular, these price differentials would vastly increase the complexity of the pricing system, and thereby also the administrative costs of its implementation. Unless the efficiency gains of the improved pricing mechanism were to outweigh the increased administrative costs, adding these sophisticated refinements would give no net gain to society.

In general, we feel that efforts to create time or space price differentials within individual irrigation systems are unlikely to be economic in most developing nations. A system of uniform prices (sometimes known as the 'postage stamp' system because it follows the post office practice of charging the same amount for all letters of a given weight, irrespective of differences in the actual costs of delivery of individual letters) would often seem satisfactory. But there are exceptions.

One exception involves situations where there are sharp seasonal differences in water scarcity. In these situations, a low price in the wet season and a higher price in the dry season could be appropriate. Another exception relates to projects with more than one source of water. Sharp differences in the marginal costs of supply associated with the different sources (for example, surface vs groundwater) could be reflected in differences in prices.

Finally, we would note that when investments are being contemplated to rehabilitate or modernise part or all of an irrigation system, opportunities exist to reflect the cost of this 'expansion of capacity' in the irrigation charges. Prior to committing public funds to such an investment, agreement could be sought from the beneficiaries to repay some or all of the new investment in the form of a special 'rehab' levy. Such a levy would be applied only to those water users whose farms are served by the rehabilitated or modernised facilities.

5.4 *Summary*

Prices are a potentially powerful tool for encouraging efficiency in the allocation and utilisation of resources. But irrigation financing policies often do little to encourage efficiency in the use of water, for the simple reason that they do not establish a true price for irrigation water. A farmer who pays for irrigation water in the form of an area-based charge still finds the marginal cost of water to be zero and thus has no incentive to economise on its use, even if the fee paid is quite high.

The failure of governments to establish water prices often reflects constraints that make implementation of such a system difficult and costly, if not technically infeasible. If the costs of implementing a price system are high, it would be economically unwise for a government to attempt to establish such a system unless the expected efficiency gains were also high. In general, such gains are likely to be high only in areas where water is very scarce, and where the government does not have an effective system for administratively rationing the water.

Volumetric measurement of water deliveries to large numbers of individual small farmers is generally costly and difficult; however, such measurements are not necessarily required for implementing a system of water prices. One possibility involves selling water volumetrically to groups of farmers who would then be responsible for its ultimate distribution among themselves. This would reduce the amount of volumetric measurements required. In some irrigation systems water prices might be established on the basis of the length of time that water is delivered, the number of irrigations that a farmer receives, or the share of the water to which an individual farmer is entitled.

Economists have written much about the value of marginal cost pricing, and how pricing systems should be 'fine tuned' to reflect differences in costs over time and space. We believe, however, that policymakers should not be overly concerned about such sophisticated details and precise determinations of the marginal cost of water, but rather focus their efforts more on major difficulties associated with the establishment of even a very simple water pricing system. Rehabilitation gives the opportunity for such a new contract between the agency and farmers with new standards of service and new levels of payment.

6

Improving investment decisions

In the previous two chapters we have considered how the financial arrangements for irrigation can affect the performance of existing irrigation systems. Better O&M and more efficient water use can improve irrigation performance, thereby increasing society's overall returns to the resources it has invested in irrigation.

But the problems of low returns to irrigation cannot entirely be addressed by efforts to improve operations. All too often the causes of poor performance can be traced to decisions made when an irrigation project was being planned, designed and constructed. These investment decisions range from design details such as the spacing, size and orientation of the channels, to the fundamental question of whether or not to undertake a proposed project.

As we indicated in Chapter 3, financial arrangements may influence the quality of these decisions. It is our purpose in this chapter to explore in more detail the prospects for using financial policies to improve irrigation investment decisions.

Before proceeding, let us remind the reader of the principal argument that we presented in Chapter 3 regarding this matter. Financial policies can be designed to increase the incentives for those who will eventually operate and use the irrigation system to evaluate the questions that are posed during the planning and design stages. But simply establishing incentives is not enough. Those facing these incentives need also to be given a voice in the decision-making process.

6.1 *Potential biases in the investment decision process*

The decision-making process for irrigation investments is often defective. The institutional structure that has been established to develop, fund and implement the construction of government irrigation projects tends to be biased in ways that lead eventually to poor irrigation performance. These biases can be seen more clearly if we first examine the nature of the process for making investment decisions in traditional communal or farmer-managed irrigation systems – that is, systems in which the government is generally not involved.

In farmer-managed systems, investment decisions – whether they be to build a new project or to undertake an expansion or modification of an existing project – are made by the collective organisation of water users that has been established to implement the project. This same group of individuals must bear all the costs of the project, including both the initial construction costs and the subsequent expenses of operation and maintenance. Because the investors and the potential water users are one and the same, it is obviously important to them that the project be carefully planned for both technical and economic feasibility. Proposals for projects that seem technically desirable but which are deemed by the users to be too expensive will not be implemented.

Of course, there is no guarantee that the investments decided upon by these communal organisations will be successful. Water users, like the rest of us, are quite capable of making mistakes! Some mistakes may be primarily technical in nature, such as when a structure fails because its design was inadequate for the actual conditions, or when the water supply proves to be inadequate to serve the entire planned irrigation area. Others are primarily economic mistakes, such as when the costs of operating and maintaining the irrigation facilities turn out to be much greater than originally anticipated. In a situation where these mistakes are severe, a project may be eventually abandoned as individual water users become reluctant to continue contributing to meet its costs.

Why do such failures occur? Fundamentally, they occur for the same reason that many business ventures in the private sector of the economy fail – because of the limitations in knowledge and in the ability to predict future outcomes. It is these limitations that make investments inherently risky ventures.

Now let us examine how the investment decision process for government irrigation projects differs from that just described for farmer-managed systems.

One key difference is that the planning and decision-making for a government irrigation project is generally done by specialists who are not, and will not become, the project's water users. Often this is a reflection of the larger scale and greater technical complexity of government irrigation projects relative to those that are built by communal groups of water users. Projects that require the construction of huge reservoirs on major rivers and the conveyance of water over long distances are generally too complex, both technically and organisationally, to be implemented by the water users.

But placing planning and investment responsibilities in the hands of specialists who have no financial stake in the ultimate performance of the irrigation system breaks a critical financial link that is inherent in the planning of farmer-managed irrigation systems. Specialisation often exists to such an extent that those involved in the initial project planning and design have nothing to do with the subsequent operation of the project – and in fact may never have any direct knowledge of how well the project actually performed after it was built. A separate cadre of specialists who often have not been involved at all in the planning and design of the project is responsible for the actual operation of the facilities.

The fact that those planning for the irrigation project have no direct financial stake in the ultimate outcome of the project already weakens the incentive to evaluate critically the prospects for the economic viability of the project. But the problem goes considerably deeper than this. Professional engineers and other specialists in an irrigation planning and design department are trained to plan and design projects. They therefore have a vested interest in seeing that new projects are brought forth for planning and implementation. The ultimate economic performance of these projects is of little immediate importance to them.

The specialised planning staff of government irrigation agencies may thus tend to become somewhat uncritical advocates for investments in new projects. But what about the position of these agencies themselves? What incentives do their leaders face? To answer this question brings us to another key difference between farmer-managed irrigation projects and government projects. With farmer-managed irrigation, the managing organisation is generally financially autonomous, while government irrigation projects are often operated by agencies that are centrally financed. The existence of central financing has important implications for the investment decision process.

Under central financing, the amount of money available to an irrigation agency for investment in new projects depends on the agency's ability to convince higher levels of government of the desirability of the investments. Surprising as it may seem, this ability is often quite independent of the actual performance of the projects.

Why should this be the case? Partly it is a matter of timing. Large projects take many years to plan and construct, and additional time before they are fully operational. It is thus not possible for the government to evaluate quickly the success of its investments in irrigation. Furthermore, determining the success of the investment even after the project has begun operating is a complex process. Project benefits, while perhaps fairly readily identifiable in general terms, may be very difficult to quantify. Thus it is often hard to determine whether the benefits are actually large enough to justify the original investment. Finally, irrigation seems often to take on the characteristic of what Jon Moris has called a 'privileged solution'.[1] By this we mean that irrigation is seen by policy makers as a self-evident solution to the problem of how to increase production. This attitude towards irrigation tends to result in uncritical or superficial evaluations of the relative costs and benefits of proposed irrigation projects.

An irrigation agency may therefore successfully promote the continued development of irrigation for a considerable period of time, even when the projects that are built are of dubious economic value to the nation. Any failure of the projects to provide a satisfactory rate of return on the government's investment becomes a burden on the general economy that is largely hidden and has no direct financial implications for the irrigation agency.

We thus find that the incentives prevailing in the context of central financing tend to create a bias at the national level in favour of irrigation investments, even if some of these investments are of doubtful economic merit. A similar type of bias occurs at the international level. As with national irrigation agencies, international lending and donor agencies also have specialised staff with vested professional interests in irrigation development. The systems of career development and professional rewards in these agencies tend to emphasise criteria such as the amount of money loaned or donated, or the number of projects undertaken. The resulting set of incentives for the staff of these agencies discourages negative recommendations for marginal projects. And generally there are no counterbalancing financial incentives to serve as a check on this bias. For example, loans made for irrigation projects by agencies such as the World Bank are almost certain to be repaid regardless of the degree of success or failure of the project, because they are guaranteed by the recipient government. Thus neither the donor agency nor the national irrigation agency has any direct financial stake in the ultimate outcome of the project.

Does this imply that these agencies are callous and unconcerned about the success of the projects that they support? Not at all. They are often very active in trying to identify and correct problems that limit the success of their projects. But the institutional context creates a set of incentives that biases decisions in the direction of being optimistic about the outcomes of new investments.

Both the specialised planners of government irrigation projects and the water users in farmer-managed projects face uncertainty and lack of knowledge about the outcome of their investment decisions. Faced with this uncertainty, both groups must weigh risks of failure against the prospects of success. The water users in farmer-managed projects are apt to give great weight to the financial risks that they personally must bear. If they are risk-averse, they may be rather conservative in their decisions, tending to invest only in the more promising projects. Specialised planners, on the other hand, do not face the same intensity of concern about the financial risks, because neither they as individuals nor the institutions in which they work need bear them. But their professional reward system may make them fairly conscious of the risk of finding too few projects to build. Thus, if they are risk-averse, they are likely to be optimistic in their projections, and tend to invest not only in the more promising projects, but also in many that are much less so.

Political pressure from potential water users is another source of bias in the investment decision process prevailing under central financing. When water users are required to pay nothing or only a small fraction of the cost of building and operating the irrigation facilities, they are likely to reap excess profits or 'economic rents' as a result of the government's investment decision. These economic rents can exist even though the project is uneconomic to the nation, simply because those who reap the benefits are not the same as those who must pay the costs. In an effort to capture this economic rent, farmers can be expected to engage in political and other activities designed to influence the government to build an irrigation project to serve them, regardless of its true economic merits from a national perspective.

Up to this point our discussion has been limited to a comparison of the investment decision process in situations of farmer-managed irrigation projects with that prevailing for government irrigation projects managed by centrally financed irrigation agencies. Two fundamental differences in these situations underlie the contrasting conclusions we have drawn. The first is the relationship of those who plan and make the investment decisions to those who use the water. In the farmer-managed systems, the planners are also the users, while in the government systems the planners are specialised professionals. The second difference relates specifically to financial arrangements. Farmer-managed organisations are financially autonomous, while the government projects were assumed to be run by centrally financed agencies. But what conclusions can we draw about the investment decision process in the case of government irrigation projects that are planned by specialised professionals but managed by a financially autonomous agency?

The first point to be made is that as in the case with the farmer-managed systems, financial autonomy for government projects has the potential to create strong incentives for the water users to evaluate carefully the likely benefits that irrigation will have for themselves, and to compare these benefits with their expected costs. Whether or not this potential will be realised, however, depends also on whether the autonomous irrigation agency is given a voice in the investment decision process. If the decisions are made by other government agencies with no input from either the water users or the irrigation agency serving them, then the characteristics of the investment decision process will be similar to the case of central financing. But if the financially autonomous irrigation agency (which will eventually have to bear the responsibility for collecting from the water users some portion of the costs of the proposed investment) is given a voice in the project planning and investment decision process, the cost–benefit calculus of the water users is more likely to be brought into the decision process, thereby improving the quality of the process.

Even when the autonomous agency has a voice in the decision process, however, the situation differs in two important ways from the case of farmer-managed irrigation systems. First, because many government projects are much larger in scope than farmer-managed projects, the users may not have a very solid basis for estimating the effects of the project. Thus their input will generally be of less help in improving the investment process than in the case of farmer-managed systems. Second, it is rare in large government projects for the water users to be responsible for the complete costs of the project. The cost–benefit calculus of the water users will, of course, reflect only the costs and benefits that they expect will accrue to themselves. If the costs only include the O&M costs, or O&M costs plus some portion of the initial capital costs, then the water users will react more favourably than they otherwise would to projects that, from a national economic perspective, may yield an unsatisfactory return.

We do not mean to imply, of course, that the willingness of users to pay the full costs of a project is a test of its economic return to the nation.

Factors such as differences between market and economic prices, the existence of non-irrigation benefits in the form of public goods (such as flood control), and regional disparities in income, may make it impossible for the water users to pay for the full costs of some projects even though those projects earn satisfactory returns.

6.2 *National case studies of irrigation investment policies*

In the previous section we presented an analysis of irrigation investment policies on the basis of economic logic. In this section we examine information on the irrigation investment experiences of three countries and evaluate the extent to which the various problems that have been identified can be linked to financial policies.

6.2.1 *The United States of America*

Although the focus of this book is on low-income countries, it is instructive to consider the financing policies used in the development of the western regions of the United States in the early part of this century.

Settlement of much of the western portion of the United States was made difficult by the aridity of the area. Irrigation came to be seen as a critical component of settlement and development efforts of the nineteenth and early twentieth centuries. The government's initial policy towards the development of irrigation was to encourage private financing. This financing, however, was generally undertaken on a speculative basis by eastern US or European capitalists, and not by the potential water users, who were often new settlers enticed to the area by the irrigation facilities. But private financing was plagued with problems. Often a considerable lag occurred between the completion of the irrigation facilities and the development of the land for irrigation. In some cases this represented a miscalculation on the part of those investing in the irrigation facilities regarding the rate of development of demand for irrigated land. In others, the owners of potentially irrigable land deliberately delayed land development in order to force the owners of the irrigation facilities into bankruptcy. Bankruptcies of irrigation companies became fairly common. Through this process, many of the irrigation facilities that had been built were acquired by the farmers at much below the original cost. Eventually, however, these problems caused potential investors to shun irrigation, and it became almost impossible to obtain funds for further irrigation development.

The lack of private funds for irrigation development led the US Congress to pass the Reclamation Act in 1902. This act provided for the establishment of a revolving fund from which loans for financing new projects could be made. Water users were expected to repay these loans over a 10-year period. The only federal subsidy was to be in the form of a zero rate of interest on the loans.

Difficulties in meeting repayment schedules under the 1902 act were soon encountered. Meanwhile, the costs of new irrigation projects continued to rise. As a result, the revolving fund (which was to have been a continuing source of financing new projects) 'failed to revolve'. Direct Congressional appropriation of funds for each individual project soon became the norm. Although repayment periods for the loans (which remained interest free) were lengthened, difficulties with repayment continued.

Responsibility for the planning and construction of new irrigation projects lay with a centrally financed irrigation agency known as the Bureau of Reclamation. Some analysts at the time criticised the Bureau for being overly optimistic in its projections of irrigation benefits and thus of the ability of farmers to repay the federal loans. One observer suggested the need for 'the Bureau of Reclamation to appraise adequately and conservatively the benefits from irrigation and to recommend to Congress only those projects for which reasonable repayment plans can be presented'.[2]

But as repayment problems continued, the concept that irrigation would require continuing government subsidies gained greater support. This idea was facilitated by arguments regarding the importance of irrigation as a means of general regional development. If irrigation, as a key to the development of the region, yielded considerable public benefits beyond its direct effects, then it could be argued that water users should not be expected to pay for its full cost.

These types of arguments continued to be elaborated, so that by 1950, the US President's Water Resources Policy Commission concluded that it would be improper to attempt to charge farmers for the full cost of water. The Commission wrote: 'but irrigation development in this country has followed a quite different course [than selling water on a commercial basis]. We have been concerned with developing the arid and semiarid West, with increasing agricultural production, with establishing independent, family-sized farms, with creating opportunities, with broadening the scope of individual property ownership.'[3]

As a result of these types of arguments, irrigation projects that clearly could not be paid for by the water users were built. A number of observers have criticised such policies, arguing that the subsidy has benefited a relatively few individuals. Furthermore, recent examinations of the costs and benefits of irrigation investments have led to the conclusion that most of the projects constructed since 1960 could not be justified in economic terms.

Although the Bureau of Reclamation which is responsible for the planning and construction of projects is a centrally financed agency, the individual irrigation projects generally are financially autonomous. But because of the large government subsidies on the construction costs of irrigation, the water users are only required to pay for a small portion of the entire cost of irrigation. A recent US government study estimated that on average, farmers served by irrigation facilities in the western United States pay approximately 19% of the total cost of irrigation. It is quite clear that under these financial arrangements, the water users have earned considerable amounts of economic rent.

Thus we see that, in this case, regional interests were successful in continuing to have irrigation investments undertaken even after it became clear that farmers could not pay for their costs. The general regional development argument for irrigation projects provided a means for justifying marginal or 'uneconomic' projects that almost certainly would not have been constructed if the users had been required to pay the full costs.

To the extent that one accepts the argument that the broad social objective of regional development should override national economic efficiency considerations in the development of irrigation, it is of course appropriate that water users should not be expected to pay for the full costs of the resulting projects. On the other hand, the use of this argument can be interpreted both as a manifestation of rent-seeking behaviour on the part of the water users, and as a justification for continued funding of a centrally financed agency, the Bureau of Reclamation, whose activities were not generating satisfactory direct economic returns.

6.2.2 *South Korea*

Modern irrigation development in Korea, which began at about the beginning of the Japanese colonial period, has been based on the Japanese institutional model of decentralised, financially autonomous irrigation organisations. The original 1906 Ordinance of Irrigation Associations authorised these organisations to finance their activities through fees imposed on the water users.

The government, apparently disappointed at the slow rate of growth of irrigation over the next decade, took steps to enhance the effectiveness of the irrigation associations. A law was passed in 1917 giving irrigation fees a legal status comparable to other taxes, thereby improving the ability of the associations to enforce the payment of the fees. Although the number of irrigation associations increased fairly rapidly in the following years, many of them encountered financial difficulties. Thus in 1927 the government passed another law that provided, for the first time, for a government subsidy on the development of new irrigation facilities.

The concept of financially autonomous irrigation associations assisted by government subsidies for the costs of capital construction has continued to the present time. Over the past 20 years, government decisions to construct complex irrigation facilities have further increased the need for financial assistance to the associations. Cost-sharing arrangements were developed to provide government subsidies for the construction of new irrigation facilities, with the exact sharing arrangements determined on the basis of the size and type of facilities built. The irrigation associations are required to incur low-interest long-term loans to cover the portion (typically about 30%) of the investment costs not covered by the government subsidy. In addition, the associations are responsible for the full costs of O&M. As a result, although there is still a large government subsidy, farmer payments for water are affected by investment decisions.

Whether this results in better investment decisions is not entirely clear. Investment decisions on new projects are made by the central government with little input from the potential water users; however, political sensitivities regarding the level of payments that farmers are required to make for irrigation may lead to a more careful scrutiny of proposed investments. On the other hand, the effectiveness of this as a mechanism of screening out uneconomic investments is offset by two factors. First, the government subsidies mean that a large portion of the capital costs need not be reflected in the irrigation fees. Second, the government's policy of using import restrictions to maintain very high rice prices has artificially increased the amount that farmers can afford to pay for irrigation fees.

For some investment decisions, such as making improvements to existing facilities, the individual Farmland Improvement Associations (FLIAs) have a greater voice. In these cases, concern over the effect that an investment decision will have on the fees that the FLIA will have to impose on the farmers should encourage a more careful weighing of the benefits and costs of proposed investments. Again, however, the fact that the central government bears much of the capital cost of the new investments reduces the FLIAs' concern about costs. A case study of one FLIA concluded that the main effect of the association's proposed investments would have been to simplify the FLIA staff's task of water distribution.[4] Many of the benefits of the investment would thus have accrued to the staff of the FLIA, while the farmers would have borne the costs. Since farmers had very little direct involvement in the decisions of the FLIA, the fact that the proposed investment would have increased water charges may have had little direct effect on the decision process of the FLIA.

This discussion emphasises the fact that financial arrangements are only one portion of the constellation of factors that affect investment decisions. A proper financial context may be an important condition for better investment decisions, but it is certainly not a sufficient condition.

6.2.3 *Philippines*

Irrigation investments in the Philippines have increased significantly since the establishment of the National Irrigation Administration (NIA) in 1964. The formal financial arrangements, specified at the time of the NIA's formation, call for the farmers to pay both for the full recurrent O&M costs and for the initial capital costs. The latter costs, however, are to be paid over a 50 year period at no interest. In recent years the NIA has charged an annual irrigation fee that, if entirely collected, would be great enough to cover the total cost of O&M plus a portion of the capital investment cost. In practice, however, the NIA's ability to collect has been weak, so that total collections have amounted to only about three-quarters of the O&M costs. The effect has been for the full capital costs and a portion of the O&M costs to have been subsidised by the government.

Until recently, the *de facto* subsidy of the full capital costs of irrigation was generally accepted by the government. Investment decisions were made as part of an overall planning process with little or no direct concern over the levels of farmer payments for irrigation services. But in the early 1980s the government, worried about how it would pay its mounting foreign debts, suggested that the NIA (a government corporation that had only a few years earlier been given a significant measure of financial autonomy) should assume responsibility for the repayment of foreign loans for irrigation investments.

This suggestion placed the question of irrigation subsidies in an entirely new light. How could the NIA, which was still having to subsidise O&M

costs from its secondary income sources, collect enough to enable it to pay off the foreign loans incurred for the construction of new projects? One possibility that the NIA considered was to refuse to undertake any new projects involving foreign loans, regardless of their inherent economic desirability. After some analysis, however, the NIA concluded that the benefits it would receive from this strategy would, at least in the short run, be more than offset by the reduction in its current income. This loss in income would have been incurred because the NIA charges a management fee for the supervision of construction and expenditures on all foreign projects that it undertakes.

The NIA's analysis of the proposed change in financial responsibilities emphasised, quite logically, the importance of the management fees that it could earn from new construction activities. But management fees can be earned on the construction of projects that are economically wasteful just as well as they can be earned on projects that are economically sound. So the role of the management fee in the financial picture of the NIA has the potential to influence investment decisions in ways unrelated to national economic efficiency criteria. This again emphasises the complex ways in which financial policies may influence investment decisions.

Some recent experiences with communal (village) irrigation projects in the Philippines also provide interesting examples of how the imposition of financial requirements on the farmers can change the ways in which projects are undertaken. The NIA often provides assistance for construction of communal projects; however, in recent years the government has made the water users responsible for the repayment of a significant portion of these construction costs. The requirement for repayment is implemented through a contractual arrangement whereby the association of water users must agree in writing to repay the costs that the NIA incurs on their behalf.

The government soon found that the water users' associations would sign such agreements only when they were given some authority to monitor and control costs incurred by the NIA. For example, they measured the amount of fuel in the government jeeps at the end and beginning of each workday in an effort to prevent unauthorised use. In some cases the farmers even kept the keys to the government jeeps at weekends! Thus the financial policies that resulted in a serious commitment by the users to repay a portion of the capital costs led to careful control by the water users over the capital expenditures. The NIA officials now had to be accountable, not just to higher level officials, but also to the water users!

6.3 *Summary*

The investment decision process in irrigation is often biased away from strict economic calculations. The existence of legitimate non-economic reasons for constructing irrigation projects is part of the reason for this; however, part of the reason can also be traced to procedures and policies that separate responsibilities for investment decisions from responsibilities for bearing the financial consequences of those decisions.

Under financial autonomy, the possibility exists of requiring that at least partial financial responsibility for new investments be passed on to the water users. But this alone will not improve investment decisions. The water users, through the financially autonomous irrigation agency, must also be given a voice in the decision process.

An examination of several national case studies of irrigation investment policies illustrates the complex nature of the environment in which financial policies operate, and the difficulties of drawing simple generalisations. But they also illustrate the key point that financial policies can have a significant effect on investment decisions. This emphasises the importance of examining financial policies from the perspective of their effect on the investment decision process.

7

Resource-mobilisation efficiency

Political leaders enjoy the fiscal activity of spending money! This creates the necessity for another, albeit less popular, fiscal activity: generating the funds to make these expenditures possible. Many alternative approaches to obtaining the needed resources exist, each with its own political advantages and disadvantages. Inherent in these fiscal decisions, therefore, are issues of political economy – a matter that we address in Chapter 12. But each alternative would also entail real welfare costs to society. These costs, as we noted in Chapter 3, include both relatively explicit administrative costs and the less obvious but still real economic distortion costs. Resource-mobilisation efficiency relates the resources acquired by the government to the total costs (including both explicit administrative costs and implicit economic distortion costs) of generating them. To the extent that the government seeks the objective of resource-mobilisation efficiency, it will attempt to minimise these costs relative to the amount of resources generated.

One common misconception about resource-mobilisation efficiency should be laid to rest immediately. It is frequently presumed that because government expenditures for the construction and operation of irrigation projects provide direct benefits to water users, resource-mobilisation efficiency implies the desirability of a user charge to capture and return to the government a portion of these benefits. No such generalisations can be made. Although, as we argue elsewhere in this book, there may well be a number of good reasons to link irrigation expenditures to a system of user fees, nothing in economic theory suggests that user fees will be the most efficient method of mobilising the resources to pay for these expenditures. The question of resource-mobilisation efficiency is fundamentally empirical in nature, and can therefore be addressed, in any given situation, only by data on actual costs.

Unfortunately, little empitical information is available on costs of alternative fiscal methods of generating resources for irrigation. In the following sections we review the principal issues that need to be considered in evaluating the resource-mobilisation efficiency of both direct and indirect financing methods, and present some limited cost data that are available.

7.1 *Administrative costs*

7.1.1 *Direct financing methods*

A decision to implement a system of user fees implies a decision to incur administrative costs that previously did not exist, and which therefore represent the incremental administrative costs of financing. These costs can be grouped into two major categories: costs for levying the fee (making assessments and billings), and costs of enforcement (including collection). The cost of gathering information for proper assessments of irrigation fees can be very high. In this respect, sophisticated systems of user fees, such as those involving metered deliveries of water to individual farmers, are likely to be much more costly than simple systems such as those based on a flat charge per hectare to all farmers. This is one of the major reasons why fees, where they exist, are so commonly based on area, especially in projects serving large numbers of small farmers. (See Chapter 9 for a fuller discussion of the relationships between the structure of the irrigation fee and the administrative costs of levying the fee.)

Collection costs can also be substantial. Developing cost-effective procedures for collecting small amounts of money from each of a large number of farmers is a major challenge to the collecting agency. Passive procedures, such as waiting for farmers to come to a central office to pay their fees, are not likely to be very effective. Repeated visits to farmers may be required. Visits may need to be timed to coincide with the sale of the irrigated crop, as a farmer may lack cash to pay the fee both before the crop is sold, and shortly thereafter. In some cases collections in kind, rather than in cash, are required or encouraged in order to improve the rate of collection; however, this entails much greater administrative and handling costs. Experience with collection in kind has shown that farmers often pay their fees with wet, low quality grain, creating costly storage and handling problems for the collecting agency. (See Chapter 11 for a fuller discussion of collection costs.)

Achieving a reasonably high rate of collection of the amounts assessed is important for resource-mobilisation efficiency, because administrative costs tend to increase more slowly than revenues as the rate of collection rises. Active collection efforts are therefore an important element of the government's enforcement procedures. But enforcement must go further than this. Unless some type of penalty is imposed on those who fail to pay, the credibility of the system of fees is likely to be undermined, leading to a general reduction in rates of collection. (See Chapter 11 for a fuller discussion of enforcement questions, including penalties for non-payment of fees.)

Meaningful information on the costs of levying and enforcing irrigation fees is usually extremely difficult to obtain. Personnel with responsibilities for these activities often also have other responsibilities for the operation of the irrigation facilities, making it difficult, if not impossible, to identify accurately the incremental costs attributable to the system of irrigation fees. Some indications of the magnitude of these costs in three different countries are given in Box 7.1.

BOX 7.1
Administrative costs of systems of user fees: examples from the Philippines, Pakistan and India

The National Irrigation Administration in the Philippines maintains data on certain costs that are explicitly incurred for the implementation of the system of irrigation fees. Data for 1984 show these costs to have been about 8% of the total amount of fee collections. But many of the NIA's employees have other primary responsibilities and also spend a portion of their time in fee collection activities, so these data probably understate the full administrative costs of implementing user fees.

In the Punjab Irrigation Department of Pakistan, about 15% of the work force are assigned as a special revenue group to assess water charges. For 1983/84, the budget for the expenditures of this group amounted to 6% of the total budget of the Irrigation Department, and was equivalent to about 10% of the total amount collected from irrigation water charges for that year. Since the actual collection of the charges (as opposed to their assessment) is undertaken by the Revenue Department rather than the Irrigation Department, the total cost associated with the collection of irrigation fees is considerably greater than the above 10% figure.[1]

In the state of Bihar, India, farmers are assessed area-based fees for the use of irrigation water. Although the funds obtained from these charges go to the general treasury of the Bihar State Government and are not directly used to finance irrigation, the irrigation department of the state is responsible for their collection, and maintains and staffs a collection unit specifically for this purpose. Information on the costs of this unit relative to the revenues collected is given in Table 7.1 below. As shown in the last column of the table, the costs of administering this area-based water charge ranged from 112% to 132% of the amounts collected for the years from 1982/83 to 1984/85. In other words, the administrative costs consistently exceed the amounts collected! This demonstrates that even when the fee structure is relatively simple, the administrative costs of implementation can be unacceptably high.[2]

Table 7.1. *Billings, collections and collection costs of irrigation fees in Bihar, India, 1982/83 to 1984/85*

Year	Current billings (million Rupees)	Total receipts from collections	Collection as % of current billings	Collection costs (million Rupees)	Collection cost as % of receipts
1982/83	71.0	53.4	75.2	60.0	112.4
1983/84	73.5	49.5	67.3	59.0	119.2
1984/85	62.4	48.3	77.4	63.7	131.9

Administrative costs may sometimes be reduced by decentralisation. This tends to place responsibility for levying and collecting fees in the hands of those who are more likely to have both direct knowledge of the situations of the individual farmers, and personal relationships useful in making the collection process more successful. In effect, decentralisation can tap an otherwise under-used resource of local knowledge, thereby lowering the overall administrative costs.

Administrative costs may be kept low by having a simple structure for the irrigation fees. For example, a flat fee based on the total area irrigated could be established, with the same rate prevailing among all projects throughout a nation. Such a fee structure could minimise administrative costs and might be the optimal approach if resource-mobilisation efficiency were the only objective.

But in many situations, resource-mobilisation efficiency will not be the only objective. The high resource-mobilisation efficiency achieved with this simple structure must be considered in conjunction with other objectives. For example, in situations where financial autonomy prevails, the water users may feel that it is more equitable for the fees to be differentiated to reflect differences in services or benefits received. (An illustration of the effect that this can have on the complexity of irrigation fees in South Korea is given in the Box 10.9 in Chapter 10.) In other words, there may be a trade-off between the resource-mobilisation efficiency objectives of a system of irrigation fees and other important objectives such as equity. As economists so often find to be true, more of one 'good thing' can only be obtained at a cost, that is, by giving up some of another 'good thing'.

7.1.2 *Indirect financing methods*

All irrigation financing methods, whether direct or indirect, will incur administrative costs. But indirect methods often have the advantage that with only a small increase in administrative costs, the amounts collected can be increased enough to provide the revenues to finance irrigation. This is likely to be true in any case where the financing mechanism itself would exist even if it were not used to provide funds for irrigation. If, for example, irrigation is to be financed through general tax revenues, these incremental costs may be essentially zero.

Using an existing framework of general taxes (whatever it may happen to be in a given country) is likely to generate additional resources at a lower administrative cost than would be incurred with any direct method of financing. But this does not guarantee that these methods are the most efficient means of mobilising resources to finance irrigation. First of all, as we discuss in the following section, indirect financing methods are likely to have greater economic distortion costs than direct methods. The second problem involves the ability not simply to generate revenues, but to generate revenues *that can finance irrigation*. Although existing tax mechanisms may be able to generate increased revenues while incurring relatively low administrative costs, it is not always feasible for the government to use the increased revenues to pay for its irrigation expenditures. Box 7.2 provides an example from Indonesia.

7.2 *Economic distortion costs*

As noted in Chapter 3, government efforts to generate revenues create economic distortion costs by causing individuals and firms in the

BOX 7.2

Resource-mobilisation efficiency and administrative costs: an example from Indonesia

For many years the Indonesian government rejected the idea of levying a direct charge on irrigation water users to generate revenues to finance any of its irrigation expenditures. Implementation of a system of irrigation fees would have required the establishment of a new administrative structure to assess, bill, collect and enforce them. When pressed by international donor agencies to establish a system of user fees to recover a portion of its irrigation expenditures, the government pointed out that the then-existing land tax, IPEDA, provided it with an indirect means of irrigation cost recovery. By increasing the productivity of the land, irrigation would result in increased revenues generated through IPEDA's administrative structure. Although the administrative costs of IPEDA might be increased somewhat by the need to reappraise newly irrigated land, it seemed likely that this incremental cost would be much less than the increased administrative costs associated with establishing an entirely new system of direct user fees.

If one could assume complete fungibility of government revenues, then this argument essentially says that the IPEDA was a more efficient method of mobilising resources to finance irrigation than a system of user fees would have been. In reality, however, the IPEDA was a tax to fund the rural development activities of local governments in Indonesia. It was not a tax specifically designed to finance new irrigation development (although a very small portion of IPEDA revenues were used by local governments for this purpose), and it was definitely not a tax to finance the recurrent costs of irrigation O&M. Increased IPEDA revenues would have permitted an increase in the rural development expenditures of local governments, but would not have reduced the need for the government to find resources to finance irrigation activities.

Recalling the distinction made in Chapter 3, we can say that in Indonesia, the IPEDA and water users' fees represented two alternative approaches to irrigation cost recovery. The real concern, however, was not for cost recovery, but for irrigation financing. And for this purpose, IPEDA was ill suited. As a cost recovery mechanism, and when measured in terms of resource-mobilisation efficiency, IPEDA appeared to be a superior mechanism. But no matter how efficient as a cost recovery mechanism for generating revenues from irrigation, it was generally unable to provide resources to finance irrigation expenditures.

economy to redirect their resources in ways that attempt to minimise the negative welfare effects of the taxes and charges. This distorts the economy away from the pattern that would maximise the net value of the goods and services produced.

Direct irrigation financing methods are unlikely to have significant economic distortion costs. Water prices may actually reduce economic distortions that would otherwise occur from providing a valuable and scarce input to the producer at a zero marginal cost. Although area-based fees lack this advantage, they are unlikely to cause economic distortions. The only exception would be if fees in existing irrigation projects were set at such a high level that the farmers decided to completely forego irrigation, resulting in the failure to utilise water that could be delivered to them at a very low marginal cost.

Indirect financing methods include taxes (both direct and implicit), inflationary financing and, in the case of financially autonomous agencies, secondary income (Fig. 3.1 of Chapter 3). Both inflationary financing and taxes are likely to have substantial economic distortion costs. Severe inflation can create large economic distortions as people try to find ways to avoid holding money, whose real value is constantly eroding. Taxes tend to distort economic decisions through their distortion of relative prices. Of the various types of taxes that may be imposed, income taxes are generally regarded as the least distorting; however, as noted in Chapter 5, they also create distortions in decisions between work and leisure. Furthermore, with complex systems of income taxes, such as, for example, in the USA, huge amounts of resources may be diverted from socially productive uses into efforts to circumvent or minimise the taxes. The opportunity cost of these resources represents part of the economic distortion costs of taxes.

A few economists have undertaken studies of the average and marginal social costs of taxation. A focus on the marginal social costs (that is, on the increase in costs that could be expected from an increase in taxes) is generally the appropriate one for our purposes, since any country will continue to have taxes regardless of how irrigation is financed. Estimates for the United States of the marginal economic distortion costs of income taxes vary considerably, generally ranging between $0.15 and $0.70 for each dollar of additional revenue generated.[3] The estimates are very sensitive to the assumed magnitude of the elasticity of supply of labour. Considerable uncertainty exists regarding the true magnitude of this elasticity, however. The studies found that using somewhat higher, but still very plausible, estimates of this elasticity leads to estimates of the marginal economic distortion costs of income taxes that exceed the revenues generated. The implication of such a situation is that in order to gain a dollar in revenues, the government would have to impose a direct cost on its citizens of one dollar, plus indirect economic distortion costs in excess of an additional dollar.

As we noted in Chapter 5, our knowledge regarding the nature and magnitude of the marginal social costs of the increase in taxation required to subsidise irrigation is quite limited. This is an area where further research is called for. But the limited information available strongly suggests that the economic distortion costs of taxation can be far from negligible. The fact that they tend to be hidden from view because they do not appear in any formal government budget does not make them any less real.

7.3 *Summary*

Any system of raising public revenues will incur costs. One objective of irrigation financing policy is to establish systems of revenue generation that are efficient in the sense that the ratio of funds raised to the costs incurred is relatively high.

Administrative costs of raising revenues to finance irrigation can be reduced if ways can be found to generate the additional revenues needed for irrigation by using existing methods of fee collection or taxation. Generally this would require the use of indirect rather than direct financing methods. But there are a number of difficulties with this approach:

(i) indirect methods are likely to have considerably higher economic distortion costs than do direct methods such as user charges;
(ii) although indirect methods may generate revenues for the government, it may be difficult to ensure that these revenues will be used to finance irrigation; and
(iii) indirect methods lack many important advantages of user charges imposed in the context of financial autonomy.

The administrative costs of user charges depend in part on the degree of complexity of the user charge system. One reason for using area-based fees rather than volumetric prices is to keep the administrative costs at a reasonable level.

Because resource-mobilisation efficiency is not the only objective, however, an approach based entirely on minimisation of administrative and economic distortion costs would not be appropriate. An irrigation agency may deliberately increase the complexity (and thus the administrative costs) of a system of user fees in order to make the farmers feel it is more equitable.

Finally, it must be remembered that for any given level of administrative and economic distortion costs, resource-mobilisation efficiency will be lower the lower the total amount of funds collected. In situations where rates of collection of irrigation fees are low because of serious enforcement problems, the total administrative costs of the system of fees can actually exceed the amount of funds collected. From a fiscal perspective, the implication is clear: the country should either give up efforts to impose an irrigation fee, or else find cost-effective ways to enhance its rates of collection.

8

The concern for equity

Irrigation (unlike rainfall) is sustained by the deployment of resources. If we want irrigation, we must therefore provide the needed resources on a continuing basis. But resources will not be provided unless someone pays for them. A key question that must be addressed in establishing irrigation financing policies is: Who should pay for these resources?

No unique, objective and universal answer exists to the question of who *should* pay for irrigation; rather, it is a question that must be answered through the political process. In any given situation these political decisions are likely to reflect perceptions of efficiency, of equity and of the positions and political power of various interest groups. In this chapter we explore the relationships between concerns for equity and policies for irrigation financing and cost recovery.

As we noted in Chapter 2, two fairly distinct types of equity concerns can be identified. The first set involves questions of the distribution of income and wealth in society. These concerns for promoting a more equal distribution of income and wealth are subsumed in the term 'vertical equity'. Considerations of vertical equity thus involve a 'macro' view of society in terms of broad social goals.

The second set of equity concerns, known as 'horizontal equity', reflects 'micro' considerations involving 'fairness' among individuals who are perceived to be equals in some sense. The underlying equity concept is that equals should be treated equally.

8.1 *Vertical equity*

8.1.1 *Common equity concerns*

Every nation tends to have specific equity concerns that reflect its unique cultural, social and political heritage. Often these concerns involve income and wealth differentials among ethnic or religious groups. It is possible, however, to identify several broad types of vertical equity concerns that are found in many nations.

One equity concern frequently encountered stems from the perception that incomes in rural areas tend to be lower than those in urban areas and that the trend over time is for this income gap to widen. Because these income differentials (which are often exacerbated by a variety of urban-biased price and investment policies) have the potential to create social and political instability, many governments are concerned to find politically acceptable ways of reducing them. One such method may involve public investments in irrigation. When this is the case, and particularly when the irrigation investment policy is deliberately designed to offset some of the urban biases associated with other policies, the idea of imposing irrigation fees on the water users is likely to be resisted. It is resisted because it appears to be in direct conflict with the policy objective of increasing incomes in the rural sector.

Another common equity concern involves regional poverty. Many nations have geographic regions in which incomes are particularly low. Often such cases are of political concern to the national government, so that fostering economic growth in these regions is seen as desirable. To the extent that irrigation development is part of a set of special economic programmes designed to achieve this, charging farmers for the irrigation services may again appear to run counter to the overall policy thrust.

Many governments also have equity concerns that focus on particularly impoverished groups within the rural sector, such as landless labourers and very small farmers. In countries where irrigation is already important, rain-fed farmers may comprise another group deemed deserving of special assistance.

8.1.2 *Equity implications of general economic policies*

Agricultural incomes are often strongly influenced by economic policies affecting the terms of trade between the agricultural and non-agricultural sectors of the economy. These include policies affecting the prices of agricultural products and inputs, exchange rates and taxes.

Whole books could be (and in many cases have been) written detailing the specific nature and effects of these policies in individual countries. An excellent discussion summarising many of these policies can be found in the World Bank's *World Development Report* of 1986. The principal observations that we wish to note here are that these policies are pervasive, and that in many cases their income distribution effects far exceed whatever effects that might be achieved through the design of policies for irrigation financing and cost recovery (Box 8.1).

8.1.3 *Equity implications of policies for property rights in land and water*

The types of economic policies considered in the previous section directly affect the distribution of current income among individuals in a nation. Property rights policies, by contrast, affect the distribution of assets among individuals. As assets can be used to generate income streams, policies affecting their distribution have important equity implications.

Even in areas lacking irrigation, policies on land tenure and land reform can have a major effect on income distribution. But they take on additional significance in irrigated areas because of the usual close linkages between land rights and effective (or *de facto*) water rights. If the sizes of landholdings in an irrigated area are very unequal, it is likely that the benefits from irrigation will also be distributed in a highly unequal fashion. In particular, we can expect that the greatest benefits would go to those farmers with the largest holdings, and who therefore presumably tend to be the richest. This represents a decrease in vertical equity, as the absolute size of the income gap between the poor and the rich farmers has widened.

But water rights do not necessarily have to be tied rigidly to land rights. One innovative example comes from Bangladesh, where the use of groundwater for irrigation is important. Rights to the groundwater are not clearly defined in a legal sense, so that possession of a well and pump gives *de facto* rights to the groundwater. In recent years there has been experimentation with a system of selling pumpsets for tubewells to groups of landless labourers who have permission to site the well either on a household plot or on someone else's field. Having thus obtained effective rights to the groundwater (in spite of their lack of land ownership rights) the group then installs and operates the well, paying for its costs by selling water to farmers with land.

BOX 8.1
Potential effects of general economic policies and irrigation financing and cost recovery policies on agricultural incomes: four examples

Thailand

Thailand, one of the world's major rice exporting nations, has long had a system of export taxation that has acted as an implicit tax on the farmers by lowering the domestic price of rice below world market levels.

Estimates of the magnitude of this tax for the period from the mid-1950s to the mid-1960s range from 80% to 85% of the farm-gate price. In other words, for every 100 *baht* received by the farmer from the sale of paddy (unhusked rice), another 80 to 85 *baht* was not received due to the government's policy of taxing rice exports. By the late 1970s the export tax had been reduced considerably, but the estimated implicit tax was still equivalent to 22% of the farm-gate price. Lower world rice prices in the early 1980s prompted the government to reduce the amount of the levy even further, so that by 1984 the implicit tax was estimated to be equivalent to only 6% of the farm-gate price. Based on 1984 prices and typical yields, this was equivalent to a tax of roughly 550 *baht* per hectare of irrigated land in 1984. Comparable estimates for earlier years would clearly be much higher.

Historically, the Thai government has not levied direct charges on farmers for irrigation services. But how does the implicit tax estimated above compare with the irrigation fees that would be needed if O&M costs were actually to be financed by fees? In budgeting funds for irrigation O&M in the early 1980s, the Thai government used, as a rough rule of thumb, an average figure of 287 *baht* per hectare. This indicates that the implicit tax that the farmers pay is greater (and in the past was far greater) than the annual cost of O&M. Thus this one major economic policy has had substantially greater equity consequences for farmers than a policy to collect water charges to finance the cost of O&M would have had.

Sri Lanka

In the years prior to 1978, the Sri Lankan government imported rice from the world market and sold it domestically at subsidised prices. The effect of this policy was to keep the domestic price of rice in Sri Lanka below world prices. It has been estimated that in the latter part of the 1970s, Sri Lankan farmers received a price for their rice that was about

30% below that which would have prevailed in the absence of the government's rice import and subsidy policy. For farmers in irrigated areas growing two crops of rice per year, this lower rice price amounted to an effective tax of approximately US$150 per hectare per year.

During this period of time the government developed, with the assistance of a World Bank loan, a portion of a large irrigation project using water from the Mahaweli River. Financial policies established for this project called for the farmers to pay an irrigation fee of $17 per hectare per year, primarily to cover operating costs. Collection of the fee proved difficult, however, with collection rates of only about 30%.

The implicit tax burden of the rice price policy was thus roughly nine times as great as the financial burden of the assessed irrigation fee. As in the previous example from Thailand, the effect of the government's rice price policy thus had much greater equity implications for the farmers than did the imposition of the user fee.

Indonesia

The Indonesian government has had significant price policies both on the major staple crop of rice and on the input of fertiliser.

Government price and market intervention policies allow it to play a significant role in determining domestic rice prices. In 1981, the effects of these government policies resulted in typical farm-gate prices estimated to be 37% below the level that would have prevailed in their absence. In the following year, however, domestic prices were maintained at a level that was probably slightly greater than would have prevailed in the absence of government interventions.

On the input side, the government has held fertiliser prices at subsidised levels. The total cost of this subsidy to the government in 1981 was approximately 314 billion *rupiah*.

Although farmers in Indonesia pay for the cost of operating and maintaining the irrigation facilities at the village level, the government has historically had a policy of charging nothing for the costs of O&M of the main system. The total cost to the government of this O&M subsidy was approximately 26 billion *rupiah*. Thus the subsidy policy for fertiliser has a total effect that is about 12 times the cost of the subsidy for irrigation O&M.

South Korea

In sharp contrast to the cases of Thailand and Sri Lanka, rice price policies in South Korea have kept domestic farm-gate paddy prices

much above world market levels. This has been achieved through restrictions and government controls on rice imports. For 1983 it has been estimated that the domestic price was 111% above the price that would prevail if restrictions on imports were eliminated. This high domestic rice price makes it much easier for farmers to pay the irrigation charge.

Irrigation is expensive in Korea. In 1983 the average cost of O&M was approximately $210 per hectare, and the average irrigation fee paid directly by the farmers was about 93% of this amount. Still, this amounts to only about one-third of a tonne of paddy per hectare, roughly equivalent to 5% of the gross production. If world prices prevailed, the same level of fees would require nearly 11% of gross production. Or looked at another way, the implicit government subsidy to the farmers resulting from the artificially high rice prices amounted to $333 per tonne. For a typical farmer in an irrigated area, this translates into a total subsidy of about $2165 per hectare, or approximately 11 times the irrigation fee that the farmer must pay. In spite of the fact that Korean financing policies call for the farmers to pay a much larger amount for irrigation than is paid in most countries in South and South-east Asia, the government's rice price policies clearly have a much greater impact on the incomes of the farmers.

Other innovative methods of allocating water rights have been devised. Under the *Pani Panchayat* programme for small-scale irrigation development in the Indian state of Maharashtra, water is allocated in proportion to the number of people in the household, rather than in proportion to a household's landholdings. Thus, for example, a household with four acres and six household members might receive enough water for three acres, while another landowner with five acres but only two household members would receive enough water for only one acre. This is an explicit attempt to use the distribution of water rights to promote vertical equity.

Water rights are sometimes bought and sold. In these cases, certain equity objectives may be achieved through the initial distribution of the rights. An example of this comes from a Nepalese communal irrigation system that was the subject of a detailed study by Martin and Yoder. The community financed the initial construction of the system by exchanging water rights for labour. Villagers who contributed labour for the construction of the system were given water rights in proportion to the amount of labour contributed. In most cases labour was contributed in proportion to the amount of land owned, so that the initial distribution of the rights tends to mirror the distribution of land. This was viewed as equitable, however, in that it also reflected both the relative contributions to the cost of constructing the system, and the relative magnitudes of the anticipated benefits. But shareholders were free to sell part of their shares to others owning land close to the system, and over time the area irrigated expanded. By purchasing water rights, owners of land outside the original system's boundaries (who of course had not contributed to the systems initial construction) were effectively contributing to the financing of the initial investment. This was perceived as an equitable way to expand the system.

8.1.4 *Vertical equity and irrigation financing policies*

8.1.4.1 *Equity implications of general subsidies*

It is often argued that subsidised irrigation promotes equity by forcing relatively rich urban taxpayers to pay for at least part of its cost, thereby reducing the financial burden that must be placed on the relatively poor farmers. This argument is superficially attractive; however, upon closer scrutiny several potential flaws appear. First, there is no guarantee that funding irrigation from general tax revenues effects a transfer from the relatively rich to the relatively poor. In any given country, the structure of the taxation system and the resulting incidence of taxation would need to be carefully examined before one could confidently draw conclusions about the nature of the transfer. In some nations, the tax system may be structured in such a way that the burden falls disproportionately on the relatively poor.

Second, implicit in the above argument is the assumption that irrigation subsidies increase taxes. This assumption is not necessarily realistic. It may be, rather, that at any given point in time, a nation's tax system generates a certain level of funds that tends to act as a general upper limit on government expenditures. If this is an accurate representation of the situation, then a decision to subsidise irrigation projects implies fewer government funds available for other projects. The subsidy thus involves an implicit transfer away from those who would have benefited if these other projects had been undertaken, and towards those who actually benefit from the irrigation subsidy. It is this implied transfer that defines the opportunity cost of the subsidy. The equity implications of the subsidy must therefore be evaluated by comparing the income and wealth positions of those who benefit from the subsidy with the positions of those who implicitly lose because of projects not undertaken. To the extent that other irrigation projects would have been undertaken, the subsidy has the effect of transferring income from presumably lower income rain-fed farmers to higher income irrigated farmers. From the viewpoint of vertical equity, this is, of course, a perverse effect.

Finally, the distribution of the subsidy among the irrigated farmers is likely to be contrary to concepts of vertical equity. To the extent that there are significant differences in the size of landholdings, we can generally expect that irrigation subsidies will benefit large farmers much more than small farmers. Some evidence for this from the United States is given in Box 8.2.

BOX 8.2

The large-farm bias of irrigation subsidies: a case study from the western United States

Subsidies on irrigation water in the United States are very high. Although the specific amounts vary by project, recent studies have estimated them to average slightly over 80% of the full project costs. These subsidies are clearly much larger than intended when they were originally established. They are, however, provided within a framework of laws designed to ensure that the subsidies would benefit primarily the small farmers.

In 1981 the US Bureau of Reclamation, which is responsible for much of the irrigation development in the western United States, estimated the distribution of these irrigation subsidies by farm size for a sample of 18 of its projects. The study found that farmers with over 1280 acres of land (who comprise only 5% of all farmers) received 50% of the total subsidies. At the other end of the scale, farmers with less than 160 acres of land accounted for 60% of the total number of farmers, but received only 11% of the total subsidies.[1] Assuming that income and farm size are highly correlated, the consequences of irrigation subsidies are contrary both to the original inventions of the policy-makers and to the general concept of vertical equity.

Of course, it might be argued that without this subsidy, the small farmers would have been squeezed out of production entirely. But if the survival of small farmers is the policy objective, the irrigation subsidy policy must be judged to be extremely inefficient, as only 11% of the amount spent reached the target group.

8.1.4.2 *Structuring user charges to promote vertical equity*

To the extent that user charges are a component of irrigation financing policies, another set of vertical equity concerns arises because of differences among the users in their income and wealth. This leads to considerations of the prospects for structuring user charges in ways that promote vertical equity among water users.

One innovative approach to this was suggested several years ago by the World Bank. The fundamental idea was to differentiate the irrigation fee according to the income level of the water users. Low income users of irrigation water would pay less for the same service than higher income users. Furthermore, water users whose incomes fell below a *critical consumption level* would not be required to pay anything for the irrigation service they received. An individual's critical consumption level is defined as the income level at which society places the same value on an additional dollar of income in the individual's hands as it does on an additional dollar of income in the hands of its government.

The rationale of the critical consumption level concept is rooted in the concept of vertical equity. Given the presumed value that society places on a redistribution of income from the rich towards the poor, it is reasonable to assume that the poorer an individual is, the higher will be the value that society places on an additional dollar of income received and retained by that individual. Society also places a positive value on additional revenues received by the public sector, since those revenues can then be used to fund socially desirable projects. For people who, because of their poverty, are able to sustain only very low levels of consumption, society values an additional dollar of income in their hands more than it values having that additional dollar received by the public sector. It would thus be contrary to society's values with respect to equity to charge people at these very low income levels for the services that are provided to them. This concept is illustrated in Fig. 8.1.

The figure compares the social value of an additional dollar of irrigation benefits captured by the government with the value of that dollar when it is retained by the water user. The social value when the dollar is captured by the government depends on the use to which the government can put it, and is independent of the individual water user's income. It is represented by the horizontal line in the diagram labelled SV-G (social value when captured by government).

But the value that society places on the dollar when it is retained by the user depends on the user's income. The lower his income, the higher is this value. This is represented by the downward sloping line in the diagram labelled SV-U (social value when retained by the user). At some income level, CCL (the critical consumption level), the social value of the dollar retained by the user will just equal the social value of the dollar when it is captured by the government. When a water user's income is below this level, society's equity goals will best be served by allowing the additional dollar of irrigation benefits to remain with the water users.

Although the critical consumption level is an attractive concept, a number of problems are likely to be encountered in efforts to incorporate it into a system of user charges. One difficulty would involve the empirical determination of the critical consumption level, since it is based on concepts that are not readily observable. But this need not be an insurmountable problem, as it certainly should be possible to develop some standard definition of an income level below which the government prefers not to remove any of the additional revenues arising from irrigation investments. It is very likely, however, that political pressures would result in setting this at a fairly high level.

Obtaining the necessary information to implement a system of user charges differentiated according to income levels would be a much more serious implementation problem. To categorise all water users according to their per capita income calls for data on both income and household

Fig. 8.1. Social values of an additional dollar of irrigation benefits.

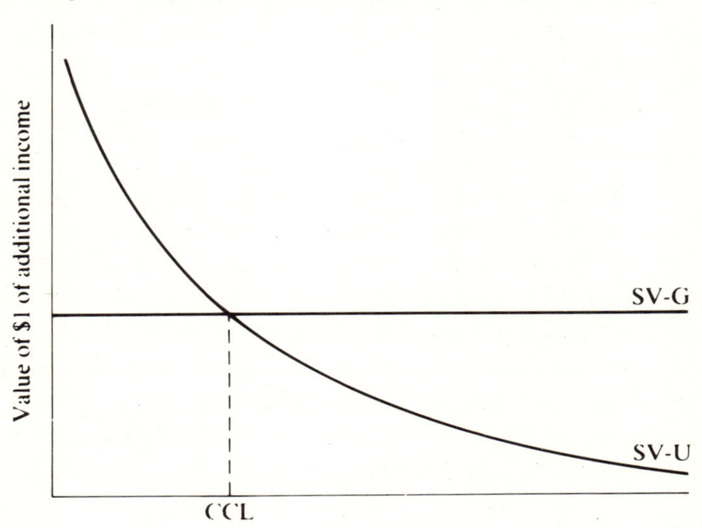

size for every water user. Reliable income data would be particularly difficult to obtain. Furthermore, this information would need to be updated periodically. Many irrigation agencies have difficulty just identifying the area that actually receives irrigation water. It is hard to imagine that these agencies would be successful in obtaining the additional information needed to implement this proposed system. Furthermore, there would be strong pressures to subvert the system. Farmers could benefit if those recording the information could be persuaded (perhaps with the help of bribes) to report a relatively low income, or a large family size. Verifying the income information, in particular, would present nearly insurmountable difficulties.

A third potential difficulty is that the water users themselves might not perceive such a system of irrigation fees to be equitable. To the extent that the users feel they are asked to pay for a service that generates additional income, they may resent a fee system that charges different amounts to individuals receiving the same service. In other words, the water users may define equity with respect to irrigation in terms of horizontal rather than vertical equity.

The danger posed by this third difficulty is that if the water users feel that the irrigation service fees are inequitable, the system of fees is unlikely to gain legitimacy. Even if the farmers initially accept the concept of vertical equity on which the fee system is based, the danger remains that over time the system will lose its legitimacy because of perceptions of widespread under-reporting of income. Once the system has lost its legitimacy in the minds of the water users, fee collections are likely to drop significantly, eventually causing the entire system of fees to collapse.

The World Bank has attempted to promote this concept for a number of years; yet it is very difficult to find examples of irrigation financing systems that differentiate fees according to some measure of income. Even within the World Bank, the approach is considered by some to be generally unworkable.

Given both this dearth of examples and the difficulties that we have discussed above, we conclude that as a general rule, irrigation financing policies should not attempt to use the structure of irrigation fees as a means of promoting society's broad income distribution objectives. At the same time, however, we recognise the desirability of providing some relief in cases of extreme poverty and in crisis situations.

One possible way to accommodate cases of extreme poverty without unduly complicating the structure of irrigation fees would be to provide for a 'subsistence exemption' that would apply to some minimal amount of every farmer's holding. Any farmer holding less than this amount of land would owe no fee at all. In principle, this approach would require no additional information to be gathered by the irrigation agency, since implementation of any system of area-based fees requires information on the total area of each farmer. It also seems likely that this approach would be seen by the water users as fair, because all users would be given the same exemption and pay the same area-based fee on all land in excess of the exemption. The major problem that can be anticipated is that farmers would have an incentive to try to increase the exemption by artificially dividing their land among family members.

To deal with crisis situations, provisions need to be made for a partial or complete forgiveness of water charges in situations of major crop loss or of crop failure due to factors outside the control of the farmer (drought, flood, typhoon, infestations of pests and diseases). Although this requires that the irrigation agency obtain additional information, a number of countries have found workable means of implementing fee systems incorporating this type of an approach.

8.2 *Horizontal equity and irrigation financing policies*

It is sometimes argued that charging users for the full cost of irrigation would be inequitable because of the existence of widespread benefits accruing to non-users of irrigation. This argument is based on the contention that both the water users and the non-user beneficiaries of irrigation should be considered as fundamentally equal. Since both groups benefit from irrigation, the requirement that one of them pay for the full cost of the irrigation system while no payment is required from the other is seen as a fundamental inequality in the treatment of equals, and therefore contrary to the concept of horizontal equity.

To evaluate this argument, we need to distinguish between direct and indirect benefits of irrigation. The application of water to land in the production of agricultural crops leads to increases in production, and thus in the net annual income derived from the land. This increase in income is the direct benefit of irrigation. It is, however, only one of two manifestations of the direct benefits of irrigation.

The other manifestation of the direct benefits of irrigation is the increase in the value of the irrigated land. In most irrigation systems, rights to irrigation water (whether explicit, or, more typically, implicit) are tied to specific parcels of land. The supply of irrigated land is thus inelastic, so that any increase in the productivity of land as a consequence of irrigation tends to cause its market value to rise. (In the jargon of economics, the increase in income becomes capitalised into the value of the land.) The increase in the value of land is a reflection of the present value of the future stream of additional annual income expected as a result of irrigation. Thus the increase in annual income and the increase in land values are two alternative indicators of one and the same thing, namely, the direct benefits of irrigation.

Typically, the water user, whose activities lead to the increase in production and annual income from the land, is considered to be the direct beneficiary of irrigation. Implicit in this view, however, is the assumption that the water user is also the owner of the land. As long as land ownership and farming are united in the same individuals, it is immaterial as to whether we consider the direct benefits of irrigation to be in the form of an increase in annual income (in which case we could say that the water user is the direct beneficiary), or in the form of an increase in land values (in which case we could say that the landowner is the direct beneficiary). Both formulations are correct because the water user and the landowner are the same individual. But when land ownership and farming are separated by tenancy arrangements, one can no longer simply equate the water users with the direct beneficiaries of irrigation. Exactly how the direct benefits of irrigation will be shared between tenants and landowners depends on complex economic, social and political relationships.

Whatever the amount of the increase in land value, it is a direct reflection of its irrigation-induced increase in productivity. It is thus appropriate to consider both water users and non-farming landowners as direct beneficiaries of irrigation. If users are expected to pay a user charge, a landowner tax on the increase in land value (either through a betterment levy or through a standard land tax) may be seen as equitable. It is true, however, that the landowners may be able to shift part of the incidence of the tax to the tenants, with the amount depending on the same complex economic, social and political relationships mentioned in the preceding paragraph. Furthermore, to the extent that farms are owner operated, the distinction between user fees and land taxes or betterment levies has no particular equity significance – they all must be paid by the same individuals.

In addition to the direct benefits of irrigation, however, irrigation also generates a variety of indirect benefits that accrue to many individuals and groups in an economy. Irrigated agriculture increases the demand for inputs such as fertiliser, pesticides and hired labour. Owners or providers of these inputs thus experience an increase in their incomes as a secondary benefit of irrigation. Likewise, the demand for marketing and processing services for the crops produced with irrigation rises, generating increases in incomes to firms providing these services.

This is illustrated in Fig. 8.2, which shows the demand and supply of marketing services for rice. When irrigation increases the production of rice, the additional rice flowing into the market creates a shift in the demand for marketing services from D_1 to D_2. Both the quantity and the price of the marketing services increase, leading to an increase in revenues ($P_2Q_2 - P_1Q_1$) and income in the marketing sector.

The expansion of firms providing inputs and marketing services generates a third round of benefits by increasing the demand for the inputs needed by these firms. In particular, owners of land in a market town that is experiencing rapid growth, due to an irrigation-induced increase in economic activity, are likely to benefit. This is so because land, being inelastic in supply in any given location, is likely to rise in price in response to the increase in demand for it. This is illustrated in Fig. 8.3, which shows the supply and demand for land in a market town. As the market grows because of the increase in economic activity, the demand for land rises from D_1 to D_2. Although the town can expand somewhat, the supply of land in a given location is relatively fixed. As a result, the

Fig. 8.2. Indirect effects of irrigation on marketing services.

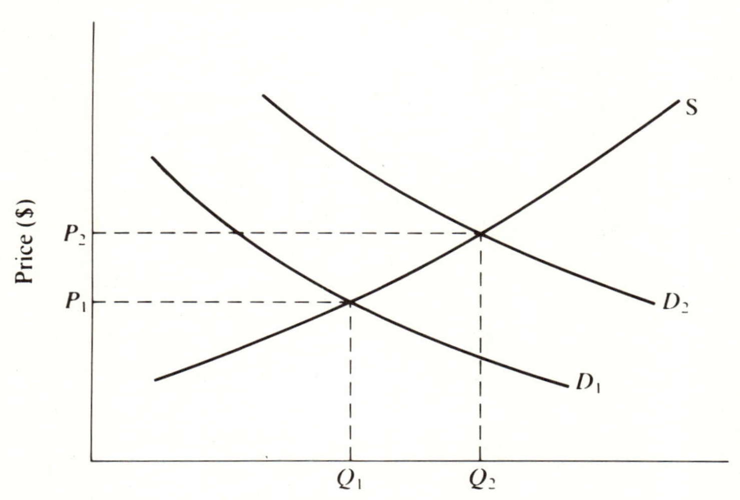

increased level of economic activity may lead to a sharp rise in land values (P_1 to P_2).

In the long run, consumers are often major indirect beneficiaries of irrigation because of the reduction in commodity prices stemming from the increased output of irrigated crops. These consumers may then redirect some of their purchasing power towards other products, thus raising the incomes of the producers of those products. Ultimately, all these indirect effects of irrigation will ripple through many sectors of the economy, often in ways that are difficult or impossible to discern and quantify.

There may also be other indirect benefits from irrigation that have some characteristics of public goods. Irrigation often increases a nation's ability to produce basic staple food crops. Many national governments view this ability as enhancing the nation's food security, with widespread socioeconomic and political benefits to all the citizens of the nation.

We have already suggested that the existence of tenancy in an irrigated area could justify, in terms of equity among the direct beneficiaries of irrigation, the imposition of both user charges and a tax on irrigated land. Does the existence of the wide variety of indirect benefits of irrigation provide a similar rationale for taxing the indirect beneficiaries?

Our first observation is a matter of practicality. Although some of the indirect beneficiaries are quite easy to identify, many of them are less

Fig. 8.3. Indirect effects of irrigation on land values in a market town.

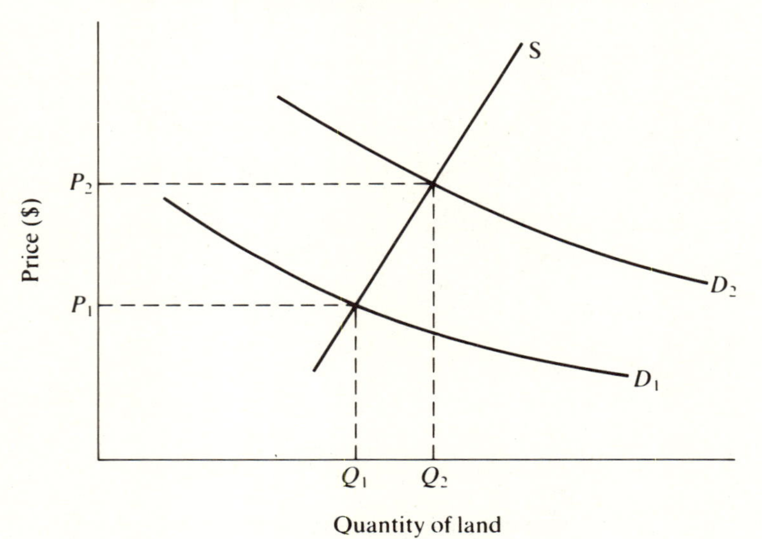

visible and are scattered throughout the economy. Practical difficulties would be encountered in an effort to establish a system of taxes for irrigation aimed at the indirect beneficiaries. When benefits are widely diffused throughout an economy, it is much easier to use the general tax system to raise revenues than to impose a beneficiary charge. The fact that in most countries a major portion of the cost of government investments in new irrigation facilities is financed from general revenues can be interpreted as political recognition of the widespread indirect benefits that accrue from irrigation. In reality, in most countries throughout the world, users pay for only a small portion of the full costs of irrigation.

Still, policy statements on cost recovery from agencies like the World Bank tend to imply that to the extent possible, water users should be expected to pay for the full costs of irrigation. In light of the equity implications of the existence of indirect benefits, how can such a position be justified?

The justification most likely to be given stems from an apparent conflict between the rationale of the above equity argument and the economist's efficiency criteria. In reacting to the above argument from the perspective of economic efficiency, an economist is likely to point out that there is nothing unique about the fact that irrigation generates indirect benefits.

Virtually all economic activity creates indirect effects. Use of fertilisers or pesticides in crop production generates many of the same indirect benefits that result from irrigation. The same is true of inputs used in any agricultural or non-agricultural activity. (In the jargon of economics, all these economic activities create pecuniary 'multiplier' effects.) Within the economic system, these effects are ultimately translated into the prices of the various goods and services that are affected. To reduce the price that a firm or a farmer must pay for any one particular input because its use creates indirect benefits, and to tax another industry in which these benefits occurred, would create a variety of distortions in market prices that would give producers and consumers incorrect signals about opportunity costs.

This point is best illustrated with an example. Assume that irrigation benefits the fertiliser industry through an increased demand for fertiliser. The effects of the increased demand will ultimately be reflected in changes in the price of fertiliser, in the quantity of its production and perhaps in the size of the capacity of the fertiliser industry. If the government were to place a tax on the fertiliser industry to recapture some of these indirect benefits of irrigation, it is likely that the price of fertiliser would rise to a higher level than would otherwise be the case. One of the many implications of this is that rain-fed farmers would have to pay a higher price for fertiliser than otherwise. The imposition of a tax (on fertiliser) that in effect forces rain-fed farmers to pay for some of the costs of irrigation is a fairly clear example of a perverse equity effect. Furthermore, by sending incorrect signals to both rain-fed and irrigated farmers about the true opportunity cost of fertiliser, this policy would cause both groups of farmers to reduce fertiliser use to levels below that which would be optimal from the perspective of society. Thus the policy creates inefficiencies in the use of fertiliser.

One further consideration about indirect benefits needs to be kept in mind. In general, except for situations characterised by significant structural underemployment of resources, secondary benefits of a given economic activity will be approximately offset by secondary costs. For example, in order to attract more labour into the processing and marketing of increased rice production, wages may rise somewhat. This will tend to raise the cost of producing other goods in the economy, thereby placing upward pressure on their prices. The result will be a negative effect on the consumers of these goods, leading to a reduction in the quantity demanded and to reduced incomes of the firms producing them. One could argue that if it is equitable to tax secondary beneficiaries of irrigation, then it would also be equitable to use these funds to reimburse those who incurred secondary costs. As a practical matter, it would be extremely difficult to undertake such a policy because secondary costs tend to be even more widely scattered throughout the economy than secondary benefits. But even if it were feasible, such an effort would create the same kind of efficiency distortions that would be encountered in the effort to tax indirect beneficiaries.

In summary, all types of economic activity create diffuse indirect benefits and costs, for which the recipients are generally not directly taxed or compensated. The equity argument for reducing user charges in consideration of the amounts of indirect irrigation benefits is thus weak unless specific features of irrigation can be identified that make it different from other inputs or from inputs in other economic activities.

We would generally argue that in terms of its economic characteristics, water as an input is not unique. However, there are two considerations related to water that might justify treating it differently from other inputs.

First, irrigation differs from most inputs in that irrigation water is typically not allocated according to price. As long as the irrigation charge takes the form of an area-based fee, the farmer's marginal cost of irrigation is unaffected by the level at which the fee is set. In these situations, reducing the fee because of the existence of indirect effects of irrigation is unlikely to have any adverse effects on the allocation of irrigation water. On the other hand, this does not justify imposing taxes on specific groups of indirect beneficiaries of irrigation, since, as we illustrated in the above example with fertiliser, it would lead to price distortions that would create both inefficiencies and other inequities. This consideration, in combination with the fact that the indirect benefits of irrigation are quite widespread, suggests a logic for subsidising irrigation from general government revenues.[2]

Second, irrigation may be considered different from inputs in other economic activities because of the nature of the resulting output. Where irrigation causes basic foodgrain prices to fall, widespread consumer benefits will exist. Furthermore, by increasing domestic capabilities to produce basic foodgrains, irrigation may be seen as creating a 'public good' in the form of enhanced national food security. Society may therefore decide that it is worthwhile to subsidise irrigation to obtain more of these particularly important and widespread indirect benefits of irrigation.

The above two considerations related to the indirect benefits of irrigation help provide both a rationale and an explanation for government subsidies for irrigation. But irrigation subsidies, by affecting the amount of economic rent earned from irrigation, lead to higher prices for the irrigated land. Based on horizontal equity considerations, it might therefore be appropriate for irrigation subsidies from general government revenues to be accompanied by the imposition of a special tax or betterment levy on the irrigated land.

In sum, horizontal equity considerations suggest that in many situations it would be appropriate to finance irrigation through a combination of user charges, land taxes or betterment levies, and general revenues. Determination of the specific proportions of the total cost of irrigation that would be financed by each of these three mechanisms would occur in the political process.

8.3 *Summary*

The first general conclusion that we wish to emphasise is the importance of placing equity concerns in their proper perspective. Irrigation raises equity issues only insofar as there are concerns about either the magnitude or distribution of irrigation benefits. But the very existence of these benefits depends on the effectiveness with which the system is built and operated. In this sense, equity concerns in irrigation financing policy must necessarily play a secondary role to efficiency concerns.

A second conclusion regarding equity and the financing of irrigation is that irrigation financing policies are not very good or powerful ways of dealing with broad income distribution concerns. The magnitude of the income distribution effects that can be achieved by modifying the amounts that farmers must pay for irrigation services is relatively small – often much smaller than the impact of other broad economic policies such as those dealing with input and output prices and with exchange rates. Given the administrative difficulties (and thus costs) of trying to achieve these relatively minor equity gains through irrigation financing policies, it would seem best to leave these policies to do their *primary jobs of mobilising resources and establishing accountability linkages that help ensure a well functioning irrigation system*. Broad equity concerns are better dealt with by other types of policies that can have a more profound impact on income distribution. It should be possible, however, to make a system of user charges more equitable by incorporating flexibility to deal with hardship cases (such as through provisions for exemptions of a minimum acreage for all farmers) and to provide for relief in times of crop failure. Financial autonomy, which helps ensure that user fees will directly benefit those paying them, also enhances the equity of user fees.

Third, equity concerns are commonly used to justify the failure to charge farmers anything for government irrigation services. But while this may give the farmers some economic benefits in the short run (as long as the irrigation facilities continue to operate reasonably well), it can generally be expected that the large and richer farmers will benefit from these subsidies far more than the small and poor farmers. This is an outcome that almost universally would be judged to be counter to society's equity objectives.

Fourth, it is commonly argued that the widespread existence of benefits from irrigation to those other than the water users makes it inequitable to concentrate on user charges for irrigation. This argument involves a number of complex considerations. Direct benefits of irrigation tend to get capitalised into the value of land, so that in situations with a significant amount of tenancy, a land tax or a betterment levy to capture a portion of the return going to landowners is likely to be seen as equitable. But the situation with respect to many of the indirect benefits of irrigation is not fundamentally different from the situation for other agricultural inputs or for other types of economic activity. These multiplier effects generally work themselves out in the price system with no strong equity implications. On the other hand, in those cases where a major indirect benefit of irrigation is in the form of reduced consumer prices for basic foodgrains, a nation may decide that it is equitable to finance irrigation partly through general tax revenues. The same conclusion holds in cases where the outputs of irrigation involve aspects of public goods, such as the enhancement of national food security through increased domestic food production. And because user charges for irrigation are not usually in the form of true water prices, these subsidies are unlikely to lead to significant distortions in the allocation of irrigation water.

Finally, we note that even when the farmers who benefit from irrigation are very poor, they are often better off than their neighbours who have no irrigation. If the farmers who have received this benefit pay for the costs of irrigation, and thus avoid a continued drain on the government budget, more funds should be available to undertake some type of development projects that may benefit those who are unable to have irrigation. This would generally be viewed as a more equitable outcome than a situation in which the government is forced to devote its resources to the continued support of the irrigated farmers.

PART III

Financial autonomy and user fees: key implementation issues

In Part II we examined the prospects for irrigation financing policies to achieve desired results relative to the five criteria of encouraging cost-effective operation and maintenance, encouraging efficient use of water, improving investment decisions, promoting efficiency in the mobilisation of resources, and promoting equity. We emphasised the key role that financial autonomy plays relative to most of these criteria.

We now turn to an examination of a number of specific policy issues associated with irrigation financing within the context of financial autonomy. We begin in Chapter 9 by considering requirements for and approaches to the establishment of financial autonomy, noting in particular the important role that user fees must play. This is followed by two chapters that focus specifically on user fees.

Implementing a system of user fees requires the establishment of the actual level of fees. Conflicting forces are encountered in this area. On the one hand, fees need to be high enough to allow the irrigation agency to perform satisfactorily the tasks of operation and maintenance. But it is unrealistic to set irrigation fees without reference to the benefits that the farmers receive. Reconciling these potentially conflicting considerations is examined in Chapter 10.

Enforcement is another difficult area that must be considered in the implementation of user fees for irrigation. If fees are assessed but cannot be collected, the system will not work. These enforcement issues are the topic of Chapter 11.

Finally, it must be recognised that formulating irrigation financing policies involves more than economic calculations. Because these policies affect the distribution of the benefits created by irrigation, they have a strong political content. Implementation of irrigation financing policies therefore requires consideration of political factors, which is the topic of Chapter 12.

9

Establishing financial autonomy

Implementation of irrigation financing policies must begin with decisions on how funds are to be generated. Are all funds to come from the government? Are users expected to provide a portion of the funds? If so, on what basis are they to be asked to pay? Are there other sources of funds that may be made available to help pay for irrigation services? These questions deal with the overall structure of irrigation financing.

In situations of central financing, the structure of financing is generally very simple: all funds come from the government, although in some cases more than one level of government may be involved. Important implementation questions exist regarding the amount of funding (dealt with in Chapter 10), but not with regard to the sources of funding.

Under financial autonomy, however, the structure of funding is more complex. Although a significant portion of an irrigation agency's revenues must come from user fees, frequently government subsidies comprise another important source of revenue. This reflects the fact that financial autonomy does not necessarily mean complete financial self-sufficiency. A third common source of funds under financial autonomy is secondary income derived either from the sale or rental of assets owned by the irrigation agency, or from special services that it renders.

In this chapter we first examine each of these three sources of financing the costs of irrigation under financial autonomy. We then consider the difficult question of how financial autonomy might be established in the common situation where the history of the nation's financial structure for irrigation has been that of central financing.

9.1 Sources of funds under financial autonomy

9.1.1 Direct financing sources: user charges and benefit taxes

For financially autonomous irrigation agencies, funds generated by direct financing methods (user charges and benefit taxes) are necessarily a major source of financing (see Fig. 3.1). User charges and benefit taxes are sometimes also found in situations of centrally financed irrigation agencies, although their role is for cost recovery rather than for financing. (Recall from Chapter 3 the distinction between financing and cost recovery.) But these levies can be structured in a variety of ways. These alternatives have implications for the efficiency with which farmers use water; for the administrative costs of implementing the system of financing; and for the users' perceptions of the equity of the financing system.

The administrative costs of implementation include the costs of levying the charge or tax and the costs of collecting it. A large component of the cost of levying the charges and taxes is associated with obtaining the information necessary to assess them. There are also billing costs resulting from the need to inform the individual farmer or landowner of the amounts assessed and of the procedures and requirements for payment. Collection costs are all those costs associated with obtaining the amounts due. Collection costs, which are sometimes very high, are discussed in some detail in Chapter 11. The following discussion thus focuses on information and billing costs.

Information costs can vary greatly depending on the structure of the specific financing method. Charges based on volumetric pricing are likely to prove very costly to implement due to the information requirements. Information would be needed for each individual farm on both the timing and the rate of flow of deliveries. In many irrigation systems, the acquisition of this information would not be feasible – which is another way of saying that it would be prohibitively expensive. Often the irrigation agency does not know from which points in the distribution network each farmer actually receives water. The impossibility of measuring the volume received in these cases is obvious. The high cost of information is certainly one reason why many irrigation agencies do not use volumetric water pricing.

In cases where the water flow is reasonably constant over time it may be possible to implement a system of water pricing that requires information only on the length of time of delivery, and not on the actual volume of water received. This is a much less demanding information requirement, and is sometimes used in small pump irrigation projects where only a few (sometimes only one or two) farmers are served by the pump at any one time. The information requirement is thus reduced, since the length of time of water delivery on any given day can be determined by the length of time that the pump operates. Furthermore, this information is relatively easy to obtain because it is available at one key point in the system where an employee (the pump operator) is present. The only other information needed is on who is receiving water each day.

Sometimes water pricing is based on a charge for each discrete irrigation. The information requirements to implement such a system are relatively low, as all that must be known is who is receiving water during each irrigation turn.

But for large systems serving many small farmers, a system of user charges based on water pricing is the exception rather than the rule. The most common type of financing methods are the area-based fees and area-based taxes, whereby the amount charged is determined on some basis in which area enters the calculation.

The simplest of all these methods is the area-based benefit tax, which relates the amount levied to the command area of the project. Each individual is charged a fixed amount per unit of land within the defined command area. The information cost for such a tax is low. All that is required is information on land holdings and on the command area of the project. No information about actual crop conditions or actual irrigation deliveries is needed.

But the very simplicity of this system is one of its weaknesses. It is not uncommon for parts of a command area to receive little or even no irrigation water. Farmers in these areas may strongly resist paying 'benefit' taxes for an irrigation system that delivers them nothing but promises. They view the tax system as inequitable. Because of these difficulties, the more complex area-based fees are generally used. For example, in Pakistan a benefit tax system based on command area was abandoned in 1979 in favour of the more complex 'crop-wise' type of fee discussed below.

The more complex the system of assessing user charges becomes, the better able it is to take into account considerations that are of importance to individual farmers, and therefore the more likely it is to be perceived as fair. However, the more complex the fee structure, the greater the information costs needed for its implementation. As in so many other areas of economics, we are faced with a trade-off. The more 'fairness' we want, the greater is the cost incurred (Fig. 9.1). Deciding what is 'best' is often a difficult task.

A first step away from the inequities of benefit taxes assessed according to the command area is a flat user fee structured simply on the basis of the area that receives irrigation water. Implementation of this type of user fee requires information on areas actually irrigated. This information, being 'decentralised' in nature, is more costly to obtain than the information needed to implement a benefit tax. Furthermore, unlike the command area, which generally remains relatively constant over time, the area irrigated by an irrigation project often varies from year to year. This creates a need for annual field assessments of the areas actually irrigated.

Many irrigation projects provide some supplemental irrigation water over a large area during the rainy season of the year, and complete irrigation to a smaller area for one or more crops during the dry season. It seems unfair to charge the same fee to farmers benefiting from dry season irrigation as is charged to those who only receive supplemental wet season irrigation. Thus the area-based fee may be structured to account for the area irrigated each season, rather than simply the geographic area irrigated. The actual fee per hectare of irrigated land may be the same for all seasons or may be differentiated by season. In the latter case, the charge for one hectare of land irrigated in the wet season is typically less than the charge for one hectare of land irrigated in the dry season. In

Fig. 9.1. Trade-off between fairness and cost in the structure of the financing method.

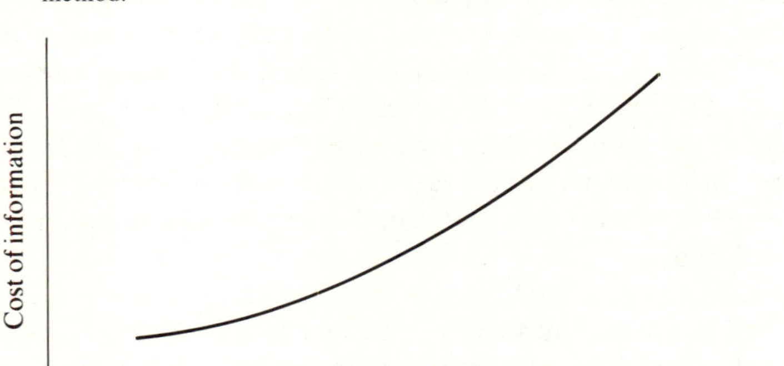

either case, this 'season-wise' area-based fee structure requires information on the areas actually irrigated for each season of every year.

An example of a season-wise fee structure comes from the Philippines. Farmers in most irrigation projects are charged a fee equal to the value of 2 *cavans* (approximately 100 kg) of paddy rice per hectare irrigated in the wet season, and 3 *cavans* per hectare irrigated in the dry season. Thus a farmer with one hectare of land would be billed for the value of 5 *cavans* if the land received irrigation water in both seasons, but only 2 *cavans* if water was only available in the wet season.

In some situations an additional level of complexity is incorporated by differentiating the fee according to the specific crop grown. These crop rates may be further differentiated by season. The information requirement for such a combined 'crop-wise' and 'season-wise' area-based fee is relatively high. Data must be collected for each farmer on the area of each crop irrigated each season. These requirements obviously place a major information cost burden on the irrigation agency.

'Crop-wise' area-based irrigation fee structures are currently found in many states in India and Pakistan. Some of these systems are quite complex, with different rates for each of a large number of crops. On the other hand, a simple crop-wise structure has recently been developed for dry season irrigation in the Philippines. In this case, where irrigation water is used mostly for rice, the fee charged for any non-rice crop is equal to 60% of the rate for irrigated rice.

Other distinctions can be incorporated into an area-based fee structure. In South Korea fees are often differentiated according to the nature of the benefits received from an irrigation project. Land that had been cultivated to rice prior to the project's construction is deemed to have benefited the least, and fees are therefore assessed at the lowest level. Previously uncultivated land that was brought into irrigated rice production as a result of the irrigation facilities is deemed to have benefited the most, and is assessed the highest fees. Another distinction often made in the structure of fees in Korea is based on the cost of providing the water. In cases where parts of the area of an irrigation association are served by a gravity system fed by a reservoir while other parts are served by pumped groundwater, differences in fees are established to reflect the higher operating costs of the pumps.

9.1.2 *Government subsidies*

Government subsidies, particularly for construction costs, are often an important source of financing for financially autonomous irrigation agencies. But they also pose a potential dilemma for these organisations, because subsidies can spell the demise of financial autonomy. The key problem is ensuring that they are compatible with the continued financial autonomy of the organisations that they subsidise. The major issues are the structure and the magnitude of the subsidies.

Structure of subsidies

Subsidies must be structured in ways ensuring that the agency's degree of success in collecting user fees continues to be reflected in the total amount of funds available to it. In the economist's 'marginal' jargon, what is needed can be stated as follows: While the subsidy can allow the agency's total budget to be larger than the amount that it collects from user fees, the amounts collected from the users must contribute, at the margin, to the size of the agency's budget. This point may be clarified by the following three examples of subsidy arrangements.

- *Example 1: Fixed percentage of new investment costs subsidised by the government.* A government could decide to pay for 70% of the investment costs of reservoir projects, and 45% of the investment costs for pump projects. By reducing the financial burden on the autonomous irrigation agencies, these subsidies would result in lower user fees, and might also affect the agencies' decisions about the types of new investments to propose. But the subsidies would in no way reduce the agencies' incentive to collect the fees assessed. The size of the subsidy is fixed regardless of the amount collected from the water users. Furthermore, the use of the subsidy is restricted to a specific purpose. For these reasons, this type of subsidy is compatible with financial autonomy.
- *Example 2: Subsidies based on a matching formula.* The government might decide to provide $1 of subsidy for each $2 of fees collected from the water users. The effect of this subsidy is similar to the previous case in that it reduces the level of user charges that must be imposed, but it does not reduce the incentive of the agency to collect from the users. In fact, this type of subsidy has the opposite effect of creating financial incentives both for the agency to collect, and for the user to pay. This occurs because the subsidy actually *increases* the marginal benefit to the agency of its collection efforts. The nature of the linkage between the subsidy and the collection of user fees is such that the size of the budget will continue to be determined by the level of

collections. This subsidy is also compatible with financial autonomy.

- *Example 3: Subsidies designed to meet budget shortfalls*. The government might agree to provide an irrigation agency with enough funds so that in total the agency would have $1.2 million to meet O&M expenditures. In such a situation, the actual amount of the subsidy would depend on the amounts collected from the water users. The more that is collected, the smaller would be the needed government subsidy. This type of subsidy obviously acts as a disincentive to high rates of payment by the farmers, and provides no incentive for the agency to be concerned with the amount of funds actually collected. It is, therefore, incompatible with financial autonomy.

Magnitude of subsidies

The total amount of government subsidies needs to remain modest relative to the overall budget of the agency. If subsidies, regardless of their structure, come to dominate the agency's budget, financial autonomy would be threatened. In order to consider the magnitude of the subsidy, a distinction is needed between the subsidies for capital costs and those for operation and maintenance costs. In many situations a financially autonomous irrigation agency is primarily responsible for operation and maintenance (O&M) activities. Assuming that *for these activities* it obtains most of its budget from user fees, it can be a financially autonomous O&M agency. Funds for construction of new or improved facilities may also be part of this same agency's budget. But if the capital costs are heavily subsidised, the agency will probably operate largely as a centrally financed agency with respect to irrigation construction (Box 9.1).

9.1.3 *Secondary income*

Financial autonomy often gives an irrigation agency enough flexibility to engage in a variety of economic activities that can generate income. In some cases, such as earning interest on funds owned by the agency, the activities are passive. In other cases the irrigation agency and its staff are actively involved. Some income-generating activities bear a fairly close relationship to irrigation, while others are totally unrelated to the agency's irrigation functions. Examples of secondary income earned by financially autonomous irrigation agencies can be found from many parts of the world (Box 9.2).

BOX 9.1
Subsidies and financial autonomy in South Korean irrigation

The Farmland Improvement Associations (FLIAs) of South Korea receive substantial government subsidies for new construction of irrigation facilities. As can be seen from Table 9.1, the amount of the subsidy received by the FLIAs depends on the type of project and on its total cost. The subsidies are thus independent of the amount of funds collected by the farmers, and are, in this sense, compatible with financial autonomy. The amounts, however, are large – ranging from 50% to 90% of the total capital cost. Furthermore, the effective amount of these subsidies is even larger because most of that portion of capital cost that is the ultimate responsibility of the FLIAs is financed by long-term government loans at subsidised interest rates. It has been estimated that the effective subsidy on the capital costs of irrigation in Korea is typically of the order of 95%. As a result, the financial autonomy of the FLIAs is very limited with respect to construction activities. In most cases the FLIAs must depend on decisions and funding from the central government. The construction itself is generally undertaken by a specialised government irrigation construction agency, the Agricultural Development Corporation.

Table 9.1. *Government subsidies for the capital costs of irrigation (%)*

Type of project	Central government subsidy	Local government subsidy	Total subsidy
Reservoirs			
Large and medium	70	0	70
Small	70	20	90
Pumping stations			
Large and medium	85	0	85
Small	70	20	90
Small weirs	70	20	90
Land consolidation			
Large scale	50	30	80
Medium scale	60	20	80
Land reclamation			
Tidal	80	0	80
Conversion to upland	50	0	50
Other	60	0	60

Although capital costs of irrigation in Korea are highly subsidised, operation and maintenance costs are largely the responsibility of the FLIAs. The two primary subsidies that exist are indirect. The first of these involves a special rate that applies to electricity used to run pumps for irrigation. This rate is approximately 43% of the rate that is charged for industrial use. For irrigation projects involving considerable amounts of pumping, this subsidy is significant. For example, for 1984 it has been estimated that in the Pyongtaek FLIA, which has a substantial area irrigated by pumps, the removal of this electricity subsidy would have required the irrigation charge to have been increased by about 15%. In projects with less pumping, the subsidy would obviously be of lesser importance. But of key importance is the fact that the magnitude of the subsidy is in no way dependent on the amount of funds collected from the water users.

The second source of indirect subsidy lies in the price of rice that farmers receive. As discussed in Box 8.1 of Chapter 8, this price is established by government policy, and is considerably higher than would prevail if the government were to allow imports of rice into the country at world market prices. Because most of the irrigated land served by the FLIAs is planted to rice, this price policy makes the financial burden of the costs of irrigation much less than would otherwise be the case. But again, the size of this indirect subsidy, while reducing the need to collect funds from the farmers, does not depend on the amounts collected. It is therefore consistent with financial autonomy, although the overall importance of this implicit subsidy to the irrigation associations has given Korean rice price policies an added political dimension.

Considering both its nature and its size, one might argue that the indirect subsidy resulting from Korea's rice price policies is not only consistent with, but also necessary to viable financial autonomy. In the absence of this high price policy, farmers might be able to pay for only a much smaller portion of the irrigation O&M costs, creating a need for large direct subsidies to the irrigation associations. These direct subsidies would tend to dominate the budgets of the associations, making them more obviously dependent on the government, and thus weakening their financial autonomy.

BOX 9.2
Examples of secondary income for financially autonomous irrigation agencies

China

Irrigation districts may undertake so-called 'sideline economic activities', which generate income that is then used to finance irrigation services. Examples of these activities include fishing, livestock and poultry production, processing of agricultural products, development of recreational uses of irrigation reservoirs, and production of non-agricultural goods in small industrial factories.

India

In some communal-type reservoir irrigation projects ('tanks') in the state of Tamil Nadu, financially autonomous irrigation associations (*ayacut* associations) are able to obtain revenues through a process of auctioning fishing rights to the tanks. Similar arrangements have been reported in Andhra Pradesh, except that the auctioning involved licences for local liquor (toddy).

Indonesia

In Indonesia, village governments often have rights to income from specified parcels of land. Village officials, including those responsible for the distribution of irrigation water within the village, are allowed to cultivate or rent out these parcels and to retain the resulting income as compensation for their services in lieu of direct payment by the water users.

Philippines

A portion of the funds financing O&M activities of the National Irrigation Administration (NIA) has come from secondary sources of income including equipment rental, interest on construction funds received but not yet spent, and a fee that the NIA charges for managing the construction of new irrigation projects.

South Korea

Secondary income is earned from sources such as interest on funds on deposit, sale of water for non-irrigation purposes, and rental of assets.

Taiwan

Irrigation associations located in urbanising areas have found that the conversion of previously irrigated land into non-agricultural uses has made some of the existing irrigation canals unnecessary. These associations have been able to sell the land on which these canals were located, invest the proceeds, and use the resulting income to finance the cost of operating and maintaining the remaining facilities.

USA

The formation of water users' organisations in irrigation projects in the western part of the United States was encouraged by government policy that gave the associations rights to certain types of secondary income, such as revenues from the leasing of government-owned project lands for grazing and farming, and profits from project hydropower plants.

The examples given in Box 9.2 indicate that secondary income is an important element in the financial picture of many financially autonomous irrigation associations. Why should this be so? The obvious advantage of secondary income is that it makes it possible for an organisation to cover its expenses for operating and maintaining the irrigation facilities while keeping the irrigation fees at a level that is acceptable to the water users (Box 9.3). It is likely that in the absence of secondary income, many financially autonomous irrigation agencies would be faced with either (1) having to raise their fees to levels that many farmers would find burdensome, (2) losing their financial viability (because of their inability to raise fees), or (3) trying to obtain subsidies from the government, which, if obtained, might cause the agencies to lose part or all of their financial autonomy.

As is the case with government subsidies, secondary income carries with it certain problems and dangers for financially autonomous irrigation agencies. The first problem is that to the extent that the agency can rely on income whose magnitude bears little or no relation to the quality of the irrigation services provided, the accountability linkages between the irrigation agency and the water users are weakened. This problem is probably not severe as long as the amount of secondary income remains considerably less than revenues from irrigation fees. But if secondary income comes to dominate as a source of revenues, this problem may become important – especially as the effects of the accountability linkages on irrigation performance are one of the primary advantages of financial autonomy.

Another potential problem with secondary income is that it may divert the attention of the irrigation agency away from its primary responsibility of providing satisfactory irrigation services. This could be severe in situations where the source of secondary income involves activities (such as raising livestock or industrial production) that bear little relationship to the operation of irrigation facilities. The problem will be minimised if the source of secondary income is an economic activity that is essentially passive in nature (such as interest earned on deposits), or that is closely linked to the responsibilities of operating the irrigation facilities.

9.2 *Moving from central financing to financial autonomy*

In the introduction to this chapter we noted that under central financing, the structure of financing was very simple, with all funding coming from the government. In the subsequent sections, we examined issues associated with each of the three sources of financing likely to exist under financial autonomy. One important implementation issue with

---

BOX 9.3

Benefits of secondary income: the South Korean FLIAs

In South Korea, the irrigation fee levied by the financially autonomous FLIAs has two distinct components: one to cover O&M costs, and the other to cover a portion of the capital cost. This fee structure has the advantage of giving the farmers, through their FLIAs, a sense of ownership rights to the irrigation facilities. But irrigation costs in Korea (both O&M and construction) are high, and it is difficult to raise the fees to a level that covers even the full cost of O&M.

So how can the farmers pay for a portion of the capital cost if the unsubsidised cost of O&M is somewhat greater than the fees? The answer lies in the secondary income of the FLIAs. In 1983, for every 100 *won* of irrigation fees collected from the farmers, an average of an additional 32 *won* was obtained from secondary income. As a result, even though irrigation fees averaged only about 93% of the O&M cost per hectare, the FLIAs were able both to cover the full cost of O&M and to meet their obligations for the repayment of loans for the specified portion of the capital cost.

---

respect to financial structure remains. In a country where irrigation operates in a condition of central financing, the government may be interested in changing the financial structure to one of financial autonomy. But to abolish suddenly a well-established system of central financing and create financial autonomy is a radical change – often too radical to be feasible. In this final section of the chapter we consider the possibilities of moving towards the establishment of financial autonomy on a more gradual basis.

The key element of financial autonomy is the reliance of the irrigation agency on funds collected from the water users or landowners. A system of user charges or benefit taxes is thus a necessary, but not sufficient condition for financial autonomy. Such systems are also found in a number of countries that lack financial autonomy. For example, in parts of South Asia user fees are actually collected by the irrigation agency; however, the funds flow into general government revenues. In these situations, some type of earmarking of the funds flowing to the government from irrigation collections could be a first step in the promotion of financial autonomy.

Several different approaches to the treatment of the earmarked funds in relation to the irrigation agency's overall budget are possible. One possibility is that after determining the total budget for the irrigation agency, the government would subtract the earmarked funds to determine the amount to be provided by the government out of general revenues. Such an arrangement, however, would effectively defeat the purpose of earmarking, and fail to lead towards any meaningful financial autonomy for the irrigation agency. A second possibility (which, in many situations, would be little different from the first) would be for the government to guarantee that the irrigation agency would never receive a budget less than the amount of funds collected from the water users.

Another and more promising possibility would be for the budget of the irrigation agency to consist of two distinct components. One component would be a general government allocation of funds to cover specified items in the budget, such as personnel, or expenditures for the maintenance of certain critical structures in the system. The second component would consist of the earmarked funds. These funds would be all the user charges or benefit taxes collected, or some previously specified portion of them. In order to promote financial autonomy, the irrigation agency should be given some discretion in making expenditure decisions regarding the use of these funds. The experience from an experiment with earmarking in Sri Lanka is discussed in Box 9.4.

BOX 9.4
An experiment with earmarking of irrigation fees: Sri Lanka

Historically, the operation and maintenance of government irrigation projects in Sri Lanka have been the responsibility of centrally financed government agencies. To the extent that irrigation fees were levied and collected, the funds became part of the government's general revenues. But by the late 1970s, very little government revenue was being generated from irrigation fees. In the 1980s, increasing concern was expressed over the mounting cost to the government of operating and maintaining irrigation projects. In 1984, the government established a new 'operation and maintenance fee' to be levied in government irrigation projects. The fee was set at an initial level of Rs 250 per hectare, which was equal to half of the estimated average annual O&M cost of projects operated by the Irrigation Department. The original plans called for the fee to rise by Rs 50 per hectare each year over a five-year period, at which point it would presumably equal (assuming no inflation or other increases in O&M costs) the full cost of irrigation O&M.

Although the agencies responsible for irrigation O&M are not necessarily directly involved in the collection of these funds, the funds are specifically earmarked for irrigation O&M. For major (larger than 80 ha) irrigation systems operated by the Irrigation Management Division and the Irrigation Department of the Ministry of Lands and Land Development, the earmarking was to be project specific. Separate accounts were to be established for each irrigation project, with funds collected from farmers in one project placed in that project's account. For each project a committee that includes the project manager and farmer representatives was supposed to make specific decisions about the use of these funds for O&M. It was hoped that these arrangements would give the farmers a greater incentive to pay the new fee.

The Irrigation Management Division's experience in the first year of fee collection was relatively encouraging. By the end of 1985, total collections of the 1984 fees amounted to slightly over 40% of the amounts due. Although this was far from outstanding, it was much better than collection rates for previous irrigation charges. But the initial success was short lived, as collection rates dropped sharply from the levels achieved for 1984. By the end of 1986, only 15% of the fees for 1985 had been collected, while collection rates for the 1986 fees stood at only 11% at the end of 1987.

What went wrong? Political difficulties and civil unrest in Sri Lanka certainly played a role in the failure of the O&M fee. But the plan also suffered from a variety of implementation problems. Some of these problems can be traced to a serious conceptual deficiency in the plan, while others reflected legal and administrative problems.

With respect to the conceptual deficiency, the plan provided for the establishment of a uniform fee throughout the nation, set at a level calculated to be 50% (for the first year) of the estimated national average cost of O&M per hectare. Given the fact that O&M costs per hectare varied substantially among individual projects, this approach to establishing the fee was fundamentally inconsistent with the concept of project-specific earmarking.

The problem becomes apparent when one considers the issue of how to allocate, among projects, the government budget covering the remaining portion (50%) of the total expenditures for O&M. From a national perspective, the costs of O&M in all projects could be fully covered only if each project were allocated enough government funds to meet the difference between the project's total costs and the funds collected from the user fee. Thus, for example, a project whose O&M costs were half the national average should receive no supplemental government budget (because the user fees would be adequate to cover in full the project's O&M costs), while a project in which the O&M costs were double the national average should receive three-quarters of its costs from the national government. But at the project level, the idea that the fee was set at an amount equivalent to half the costs of O&M led to the expectation that the fees collected would be matched by government funds. From the perspective of the water users in a given project, what was the significance of earmarking their user fees for use in their project if the existence of these funds simply allowed the government to reallocate its budgetary support to other projects? To be meaningful, project-specific earmarking of user fees would have required project-specific agreements about the government's portion of the O&M budget, either in terms of amounts of funds to be provided or in terms of specific responsibilities to be undertaken by the government. But the plan failed to provide for this.

Sri Lanka's efforts to establish a user fee encountered a variety of other implementation problems stemming from legal and administrative difficulties. Legal challenges were raised to some of the provisions. The procedures for making expenditure decisions on the funds collected from the users were cumbersome, creating considerable delays between the time that the fees were paid and the time that O&M expenditures based on these funds were made. Enforcement turned out to be a serious problem. The number of court cases that emerged was so great that the courts were unable to process them promptly. For the cases that were handled, the court procedures were slow, and a number of farmers won their cases. Furthermore, even when the courts ruled against the farmers, no penalty, other than an order to pay the fee, was imposed.

Faced with these difficulties, the Irrigation Management Division decided not to try to raise the annual fee in accordance with the original plan, focusing instead on trying to increase the rates of fee collection. But as the data on rates of collection indicate, this was not very successful. When viewed from the perspectives of the farmers, the sharp drop in the rates of collection after the first year is not particularly surprising. The delay in expenditures meant that farmers who had paid their 1984 fees saw no immediate benefit in terms of the operation or maintenance of their irrigation facilities. At the same time, they could observe that their neighbours who had not paid the fee did not suffer any serious consequences.

This experience demonstrates how difficult it can be to move even in a modest way toward financial autonomy when the necessary system of user fees is not already well established. One lesson is that in moving toward financial autonomy, it is important that the level at which this autonomy is to exist needs to be clear and consistent. In the Sri Lankan effort, some elements of autonomy were implied at the project level (project-specific earmarking of fees), while others were at the national level (nationally uniform rates of assessments, and a national budget to pay for the portion of O&M costs not covered by fees). This led to a fundamental inconsistency between the responsibilities of the national irrigation agency and the perceptions of the water users in individual projects. This experience also provides a lesson on the importance of working out all the detailed implementation procedures prior to undertaking such an initiative. Much of the confusion, uncertainty and delays in expenditures that emerged, as well as some of the unfavourable (to the government) court rulings could probably have been avoided if the new fee structure had not been implemented before all of the legal and administrative details had been thoroughly worked out. Finally, this experience demonstrates the importance of enforcement. Unless the system can keep the 'free riders' down to a small number, they can quickly undermine the entire foundation of the system of irrigation fees.

Earmarking of funds collected from the water users can be expected to have the greatest impact in the direction of financial autonomy if the irrigation agency itself is responsible for the collection of revenues. This is because the links that the earmarking creates between the agency's budget and the amount of funds collected gives it a financial interest in collections. The impetus of earmarking towards financial autonomy is likely to be weaker if the funds are collected by some other agency (such as a government revenue department) that lacks this financial interest.

In the absence of any system of payments for irrigation by users or landowners, earmarking of certain funds for irrigation O&M may be possible, but it will be very difficult to establish even the beginnings of financial autonomy. In this situation, earmarking may help protect the irrigation agency from the vagaries of the government budget process by ensuring some minimum level of funding; however, it does not give the agency any control over the amount of funds it receives (see Box 9.5 with the example from Indonesia). As a result, whether service is good or poor has little bearing on the determination of the size of the agency's budget.

Earmarking of funds collected from water users can be a very powerful step in the direction of financial autonomy. Ultimately, there may be little difference between a situation where earmarked funds from user fees flow to the irrigation agency through the general revenues, and one where the funds from user fees are retained directly by the irrigation agency.

The critical obstacle to increased financial autonomy at this stage is likely to be the amount of government budget subsidy for O&M. To deal with this problem, a carefully planned programme for the reduction of these subsidies needs to be worked out. It is important that this reduction not be linked to the degree of success in collecting fees from the water users. Otherwise, it will act as a disincentive to the collection and payment of the water charge.

The transition from central financing to financial autonomy is difficult but not impossible. One nation which has recently had experience in creating financial autonomy is the Philippines. This experience is discussed in Box 9.6.

9.3 *Summary*

Although centrally financed irrigation agencies receive all their funds from the government, financially autonomous irrigation agencies

---

BOX 9.5

Earmarking land tax revenues for irrigation: an example from Indonesia

Indonesia has had no system of water users' fees to cover the irrigation O&M costs incurred by either the national irrigation agency, the Directorate-General for Water Resources Development (DGWRD), or the Provincial Departments of Public Works, which have O&M responsibilities for government irrigation projects. These agencies are thus centrally financed.

Over the years, these agencies have faced difficulties with the low level at which the government has funded their operations. External donors, such as the World Bank, have felt the need for increased funding for O&M in order to prevent the rapid deterioration of new irrigation facilities. The difficult question has been how to provide for this.

Given the lack of any direct charges on water users, attention was focused in the early 1980s on Indonesia's land tax, IPEDA. Because the tax was assessed on the basis of productivity of the land, the revenues that the government received from IPEDA should have increased as a result of irrigation. This provided the logic for trying to earmark a portion of this tax for irrigation O&M.

The suggestion for earmarking was not easy to implement, however. IPEDA had traditionally been used as a rural development tax, with local discretion over the allocation of the funds. The IPEDA revenues tended to be used to develop new projects, not to operate and maintain existing ones. As a result, there was considerable resistance to the idea of earmarking a portion of the tax for irrigation O&M. Eventually, however, a proposal to do so emerged, along with a change that eliminated the IPEDA in favour of a more broad-based real estate tax (known as the Land and Building Tax).

The implementation of such a proposal could help ensure that DGWRD and the Provincial Departments of Public Works have funding that is not subject to the whims of the government's annual budget allocation process. It thus might provide a more stable source of financing. But the mechanism would give them no real degree of financial autonomy. They would still be dependent on their stated share of the tax, over which they have no real control. In contrast to the situation with true financial autonomy, the irrigation agencies would have little ability to affect the amount of funds that they receive by improving their performance.

BOX 9.6
Developing financial autonomy: the case of the Philippines

The National Irrigation Administration (NIA) of the Philippines was established in 1964 as a semi-autonomous government corporation. For the first decade of its existence, however, it operated primarily as a centrally financed government bureau. Operating funds came from annual appropriations from the national government, and funds from the NIA's collection of irrigation fees were not retained by it, but rather had to be remitted to the national government.

The situation changed with the amendment of the NIA's charter in 1974. This amendment provided the basis for the transformation of the NIA into a financially autonomous agency with respect to its operation and maintenance activities. Under the new arrangement, the NIA was allowed to retain all the funds that it collected from the water users. Recognising that the NIA's revenues from fee collections would not immediately be enough to cover the costs of O&M, the government included a provision for a subsidy for O&M that would gradually decrease to zero at the end of a five-year period.

The transition to a financially autonomous agency was not easy for the NIA. Shortages of funds forced it to lay off some of its field staff. Further reductions in the size of the field staff were achieved through attrition. In many small systems, where the NIA's management structure resulted in a high administrative cost per hectare of land irrigated, the NIA attempted to reduce costs by encouraging the development of water users' associations to which partial management responsibilities could be delegated. The underlying presumption was that these associations would be able to undertake certain management responsibilities at a lower cost than the NIA could achieve.

The process of transition toward financial autonomy was made somewhat easier by the fact that the NIA had access to a considerable amount of secondary income that it was able to use to subsidise its O&M activities. The primary source of this secondary income was from management fees that it charged for the supervision of construction of new irrigation projects funded through foreign borrowing. Without this source of funds, it is doubtful if the NIA could have continued to operate and maintain the irrigation facilities in a satisfactory manner. The primary obstacle that the NIA, even today, must overcome is that of low collection rates. Financial autonomy has given it a much greater incentive to improve fee collection,

and it has made considerable efforts in this direction. But until collection rates can be increased even more, the NIA is likely to continue to face the need to fund a significant portion of its O&M costs from other activities.

In considering the lessons of the Philippine experience for the development of financial autonomy, two points stand out. First, a system of fees already existed prior to the attempt to make the NIA financially autonomous. Second, the NIA's legal structure as a government corporate entity greatly facilitated the development of financial autonomy. Although a change in its charter was required, it was undoubtedly easier to allow the NIA to retain the fees it collected because it was a government corporation than it would have been had the NIA been a regular government line agency.

---

often have three sources of funds: user charges or benefit taxes, government subsidies and secondary income. Of these three sources, however, only the first is necessary for the continuation of financial autonomy.

Direct charges for irrigation (user fees or benefit taxes) are thus critical to the financial survival of autonomous irrigation agencies; however, they can also be costly to administer. It is therefore important that careful consideration be given to the trade-offs between a simple structure that can be implemented at a low cost, and a more complex structure that may seem more equitable (and therefore more acceptable) to the water users.

Government subsidies will continue to exist, albeit we anticipate at a declining proportion of total costs. Government subsidies can be consistent with financial autonomy, as long as they are structured so that they do not reduce the effect on the agency's budget of its efforts to collect the direct charges, and as long as their magnitude remains modest in relationship to the agency's overall budget.

Secondary income earned from other sources can supplement the total resources available to the irrigation agency, making it possible to operate at a lower level of direct charges than would otherwise be possible. Many financially autonomous irrigation agencies have sources of secondary income; however, if the amounts become too large, there is a danger that the managers of the agency will become so preoccupied with earning secondary income that operation of the irrigation facilities could suffer.

Moving from central financing to financial autonomy can be a radical and difficult institutional change. Earmarking of funds for use in irrigation operation and maintenance is one possible approach to a more gradual transition to financial autonomy. Giving irrigation agencies special administrative status outside of regular government line ministries can also facilitate a move toward financial autonomy.

10

Setting irrigation fees: reconciling the need for funds with farmers' ability to pay

10.1 *Introduction*

'User fees for irrigation are fine in principle, but irrigation is expensive. Is it realistic to expect that low-income farmers can afford to pay for the costs of irrigation?' This fundamental question, which is commonly raised in deliberations on national policies for irrigation financing, is the focus of this chapter. Before proceeding further, however, we need to clarify what we mean by the term 'afford'. We need to distinguish between a narrow, strictly economic meaning of the word and a broader meaning that also incorporates political concerns.

In a strictly economic sense, a water user can 'afford' an irrigation fee as long as that fee is smaller than his or her additional net income (prior to paying the fee) that is attributable to irrigation. In other words, as long as the irrigation fee does not leave the water user with less net income than would have been received in the absence of irrigation, the fee is affordable in an economic sense. The evidence from private irrigation is that in this sense of the word, poor farmers often can afford to pay quite large amounts for irrigation (Box 10.1).

But in a broader political context, the term 'afford' is not likely to be defined in this strict economic sense. Rather, it is more likely to be defined in terms of the amounts that governments are willing to ask the water users to pay. Even with good irrigation facilities, irrigated farmers may be poor relative to some national poverty standard. This may lead a government to argue that although irrigation has improved their incomes, water users still cannot afford to pay for its costs. This political definition of 'afford' implies a concern either for equity, which we have dealt with in Chapter 8, or for practical politics.

In this chapter we focus our attention on the narrower economic meaning of 'afford', exploring the relationships between the costs of irrigation and the size of the irrigation benefits.

10.2 *Cost-based vs benefit-based irrigation fees*

Within the literature on irrigation financing one finds some controversy over the question of whether the fees should be based on the costs of providing the irrigation services, or on the basis of the benefits received.

Some economists argue that water charges established on the basis of benefits rather than costs would be inefficient. The presumption underlying this argument is that the irrigation fee is in the form of a true water price. If those with high total benefits from irrigation were charged a higher price than those whose total benefits were low, then the two groups of farmers would face different marginal costs for water. Farmers

---

BOX 10.1

Can poor farmers afford to pay for irrigation?

Information on communal and private irrigation systems in various countries in Asia shows that even very poor farmers often pay quite large amounts for good quality irrigation services.

- **In Bangladesh, it is not uncommon for a farmer to agree to pay 25% of his dry season irrigated rice crop to the owner of a nearby tubewell who supplies the water.**
- **Studies of farmer-managed irrigation systems in Nepal have revealed that farmers contribute large amounts of cash and labour to pay the annual costs of operation and maintenance. For example, in six hill systems studied in detail, the average annual labour contribution was 68 man-days per hectare. In one 35-ha system annual labour contributions were appoximately 50 man-days per hectare, while cash assessments averaged about 350 Rupees per hectare, which, at the local wage rate of 10 Rupees per day, is equivalent to over one man-month of labour.**

These observations lead to the conclusion that although the payments are large, the benefits that farmers perceive they are receiving from the irrigation services must be even larger. Although many of these water users are very poor in an absolute sense, they can afford to pay these costs of irrigation because they are still better off than they would be if they had no access to irrigation.

---

within each group would adjust their purchase of water to the point where the marginal benefit was just equal to the price paid. Thus the marginal benefit of a unit of water would be higher for those with high total benefits than for those receiving lower benefits, indicating that a reallocation of water from the latter to the former would result in an overall increase in the value created by the irrigation water. The validity of this argument is, of course, limited to situations where true water pricing prevails. As we have noted elsewhere, this is the rare exception rather than the rule for irrigation systems in developing nations. Furthermore, the argument is based on the assumption that the charge would be differentiated according to the level of benefits received. But in many cases a uniform benefit-based charge may be established, taking into account the benefits received by the majority of the water users.

Other economists, noting that there is often a considerable amount of inefficiency and corruption in the process of designing, constructing and operating irrigation systems, argue that it would be inequitable to ask the water users to pay for these 'leakages'. They thus suggest that irrigation fees should be based on the benefits received by the users, rather than on the costs incurred in providing the irrigation facilities.

In practice, both costs and benefits need to be considered in the establishment of irrigation fees. It is clearly unreasonable to ask users to pay fees for irrigation service that exceed the benefits they receive from that service. An irrigation agency wishing to establish a user fee must therefore be concerned with the nature and magnitude of these benefits. We examine questions related to benefits in Section 10.3. On the other hand, costs cannot be totally ignored, especially in the case of financially autonomous irrigation agencies. This suggests the need to allow users to have a voice in expenditure decisions – a point that we have made in Chapters 4 (regarding O&M) and 6 (regarding capital costs), and to which we return in Section 10.4 of this chapter.

10.3 *Ability of farmers to pay for irrigation*

Imagine a farmer making a decision about the purchase of a small pump to irrigate his land from a nearby stream. If he has plenty of money and thinks that a pump will enhance his prestige among his neighbours, he might decide to purchase it without considering its expected economic return. But more likely, his decision about the purchase of the pump would be based on his evaluation of the economic benefits that the pump would generate to him, relative to its cost to him. In other words, he would only decide to buy the pump if he expected that he could 'afford' it in the economic sense of being able to pay for it from the increased income it would generate. If, after deciding to purchase the pump, he finds that he cannot 'afford' it in this sense, then the pump is not living up to its expectations. Possible reasons for this could be unrealistic expectations, unexpected changes in market prices (for example, a sharp drop in the price of the irrigated crop), or poor irrigation management practices by the farmer.

For many public irrigation projects, an investment is made only after analysis has shown that a proposed project can be expected to provide a satisfactory economic return. For goods or services where market prices do not reflect social opportunity costs, shadow prices (estimates of their opportunity costs to society) are used in the calculation of the economic return. Where these shadow prices differ sharply from market prices, it is possible that the water users could not afford to pay for the total cost of irrigation, even though the investment shows a favourable economic return to the nation. But barring this possibility, a failure of the farmers to be able to afford to pay the costs of irrigation implies about the same things that would be implied in the case of an individual investment decision: unrealistic expectations, unexpected changes in market prices, or poor irrigation management practices.

Of course, with public irrigation projects there is greater scope for unrealistic expectations in the investment decision process. This reflects not only the greater uncertainty about outcomes, associated with the larger scale of government projects, but also the tendency (discussed in Chapter 6) for public investment decisions to be biased toward projects of dubious economic merit due to weak financial linkages between the investment decision process and the outcomes of the project. And with public irrigation projects, the scope for poor management practices has expanded beyond the individual farmer's management of irrigation, and now also includes the entire irrigation agency responsible for operating the public facilities.

We should also mention one significant administrative complication that may make it difficult for a public agency to collect the cost of irrigation services from the water users, even if they can afford to pay for it. Typically, the per hectare irrigation benefits received by individual farmers within a single irrigation project will vary considerably both from year to year and, for any given year, from place to place within the project. It is generally not practical for a public agency to determine what benefits each individual farmer receives, and to adjust the irrigation charges accordingly. Rather, some general fee structure must be established and applied to all the water users. If the agency establishing the charge wants to be sure that it does not impose an unreasonable burden on any of the water users, it may be forced to set the charge at a relatively low level. As a result, the charge may not generate enough funds to cover the full costs of irrigation (Box 10.2).

Finally, we need to mention that because of other objectives such as food security, investments in irrigation are sometimes made in spite of very low projected rates of return. In these cases it would clearly be unrealistic to expect that the water users would be able to pay for the full costs of irrigation.

The above discussion has focused on the full costs of irrigation. But what if we consider only the costs of operation and maintenance? Since these are usually much smaller than the total costs of irrigation, it seems reasonable to expect that farmers would be able to pay for them. This, in

---

BOX 10.2

Setting a uniform fee in relation to the ability of most farmers to pay: an illustration of the effect on rates of cost recovery

Even in cases where irrigation projects are generating significant net benefits, uncertainty about their magnitude, and a recognition that they are unevenly distributed both over time and space may make it impossible for a government irrigation agency to set a fee that would be high enough to recover the full cost of irrigation. To illustrate the problem, we will consider a simple example of a project with a favourable benefit–cost ratio of 1.5 in which all of the benefits are in the form of increased crop incomes, and there are no differences between economic and financial costs and benefits. We assume that the average annual benefit per hectare is $150. The total annualised cost of the project is thus $100 per hectare.

If the benefits were distributed uniformly throughout the project area, then a fee set at $100 per hectare, or 67% of the net benefits, would fully recover the costs of the project. But a government irrigation agency, uncertain about the precise level of benefits actually achieved, as well as about variability from year to year, is unlikely to be willing to set the fee at such a high level. A more realistic maximum target level of fees would be 50% of the estimated average benefits. This causes the level of cost recovery to drop to 75%.

In reality, the benefits are not likely to be distributed uniformly throughout the project area. For purposes of illustration we assume that the distribution of the benefits is such that 90% of the farmers receive benefits of at least 40% of the average, while the remaining 10% of the farmers receive benefits of less than this amount, including a small number who receive no benefits.

As long as some farmers receive no benefits from irrigation, there is obviously no uniform fee that can be established that would not exceed the ability of some farmers to pay. Recognising this fact, the government might establish a policy goal of setting the fee at a level that, for at least 90% of the farmers, would not take more than half the expected benefits. Under the assumptions given above, this implies a maximum fee of $30 per hectare, giving a cost recovery rate of only 30%.

It is obvious that most of the farmers would have the ability to pay considerably more than this amount. But the more the fee is raised toward the amount necessary to obtain full cost recovery, the larger will be the number of farmers for whom the fee exceeds the benefits received. Given the strong negative equity implications of a fee that exceeds benefits, one can expect political pressures to keep the level of the fees relatively low.

---

fact, is the conclusion of some recent studies that estimated the relationships between irrigation fees and typical irrigation benefits in five Asian countries (Box 10.3).

The results of the five-nation study cited in Box 10.3 support a conclusion that seems intuitively valid: irrigation that is functioning in a reasonably satisfactory manner can be expected to generate enough benefits to easily cover the costs of operation and maintenance. Except perhaps in cases of some pump irrigation projects, which could have very high O&M costs, this conclusion should hold.

For a government wishing to establish financially autonomous irrigation agencies, an important policy implication of the above conclusion is that a minimum target of having the agencies be fully responsible for O&M costs is, under normal circumstances, likely to be reasonable. But a second implication is that it is probably unrealistic for these agencies to be responsible for the full amount of capital costs.

10.4 *The need for funds: budgeting for O&M*

Thus far in this chapter there has been no discussion of the nature of O&M costs. What types and amounts of expenditures are to be

---

BOX 10.3

Ability of water users to pay for public irrigation

In a recent study of five Asian countries, it was concluded that as long as irrigation facilities were performing in a reasonably satisfactory fashion, the direct benefits accruing to the farmers would generally be large enough to enable the farmers to pay for the full cost of O&M. It was estimated that if current policies in these five countries were modified to incorporate a user fee set at a level to cover the full cost of O&M, the fee would take 7–36% of the net benefits received (Table 10.1).

But the study also concluded that in most cases the farmers could not realistically be expected to pay, in addition, for more than a portion of the capital costs. To attempt to recover both O&M costs and the full capital costs would often require user charges considerably in excess of the average net benefits of irrigation. One interpretation of these results is that many irrigation investments (particularly those involving fairly high capital costs) are not earning a satisfactory economic return. This could reflect either defects in the investment decision process (see Chapter 6) or unsatisfactory operation and maintenance of the facilities (see Chapter 4).

The figures in Table 10.1 on the percentage of benefits needed to pay for the costs of irrigation do not incorporate consideration of the costs of collecting user charges. If these costs are high, the percentage of benefits needed to pay for the costs of irrigation would rise accordingly.

In this regard, the failure to collect amounts owed is a particular problem. The same study also estimated average rates of collection of the

Table 10.1. *Estimated percentages of a typical water user's net irrigation benefits needed to pay irrigation fees under alternative financing policies in five Asian countries*

	1984 policy	1984 policy modified to set user fees equal to O&M costs	1984 policy modified to set user fees equal to O&M plus capital cost: Moderate capital costs	1984 policy modified to set user fees equal to O&M plus capital cost: High capital costs
Indonesia	8–21	10–27	56–154	114–313
Korea	26–33	27–36	141–183	203–264
Nepal	5	10	74	122
Philippines	10	7	43	98
Thailand	9	31	155	279

amounts due from the farmers in three countries (Korea, Nepal and the Philippines) that had systems of irrigation charges imposed on individual farmers. In the case of Korea, the rate was 98%. By contrast, the rate of collection in Nepal was estimated to be only approximately 20%. The Philippine collection rate was at an intermediate level, at about 62%.

Given these figures, it is a simple matter to calculate the irrigation charges per hectare necessary to cover O&M costs in full from the funds collected (assuming that the increase in fees would not affect the actual collection rates). In the case of Nepal, with only a 20% collection rate, the charge per hectare would need to be five times as high as would be necessary if the collection rate were 100%. This would imply that to cover the costs of O&M in full, the fee would take 50% of the net benefits of irrigation, rather than the 10% figure shown in Table 10.1. Similarly, the figure for the Philippines would rise to 11% from the figure of 7% in Table 10.1.

---

incurred for O&M? This is an important question because the answer to it affects both the quality of O&M and the amount that a financially autonomous irrigation agency will need to collect from the water users.

As we noted in Chapter 4, questions of the types and amounts of expenditures to incur for O&M are frequently treated as technical matters to be decided without reference to the water users. The task of setting irrigation fees thus becomes one of establishing a fee structure that will generate the 'required' amount of funds.

But water users, as a group, are effectively consumers of irrigation services. As such, they have important perspectives on the relationship between the quality and the cost of these services. Certain expenditures deemed desirable by the technical staff of an irrigation agency may be seen by the water users to be too costly relative to their contribution to the quality of the services provided. Or the users may find that the irrigation agency's cost of undertaking certain components of O&M are so high that they prefer to assume direct responsibility for implementing these components themselves (Box 10.4).

That water users who are being asked to pay for the costs of O&M have a legitimate interest in decisions about the nature and magnitude of O&M expenditures does not deny the role of the technical staff of the irrigation agency in these decisions. Clearly, some aspects of O&M may involve technical considerations about which the water users may have little

---

BOX 10.4

The role of water users in making expenditure decisions: an example from the Philippines[1]

Talaksan is a small communal pump irrigation system, serving about 38 ha, and operated by the farmers through an Irrigators' Service Association (ISA). Engineers from a government agency, the Farm Systems Development Corporation (FSDC), provided technical assistance in the design and construction of the system. Under government policy, construction costs incurred by the government were financed through a loan to the ISA carrying an interest rate of 12%. The original estimate of the cost of construction was approximately $7000; however, the loan, based on actual construction costs, was for only $4100. How was such a large reduction in costs accomplished?

In part the saving was due to the fact that the farmers contributed certain construction materials. But a significant portion of the savings was the result of a farmer-initiated elimination of an elaborate stilling basin from the plan. This change was agreed to by the engineers. Additional costs could have been saved if the engineers had agreed to another of the farmers' requests: the elimination of a large concrete end check designed to control water at the end of one canal. The end check was constructed in spite of the farmers' reservations, and they had soon by-passed it in order to extend the canal beyond its original designed length.

Other changes were made by the farmers. Although the total area served by the system was almost exactly the same as that in the original plan, the total length of channels constructed (mostly with the farmers' labour) was 2.7 km, compared with the original plan of 1.5 km.

The farmers thus significantly changed both the amount and the nature of the expenditures incurred for the construction of this project. These changes were based on their perspectives of both what was important relative to the costs, and what they could more easily contribute (i.e. finance) in kind rather than in cash.

---

direct knowledge or information. In the end, what is needed is an integration of the perspectives of the technical people with those of the water users, so that a reasonable balance between expenditures for O&M and the benefits of those expenditures is obtained. It is in this sense that the 'need' for funds for O&M and the ability of the users to provide these funds are interdependent and should be determined jointly.

The example of the input of water users in expenditure decisions noted in Box 10.4 represents an unusually large role for the water users. The intensity with which the users were involved in these decisions reflects two conditions. First, the water users were obligated to pay, through the amortisation of a loan, the cost of the expenditures – i.e. financial autonomy prevailed. Second, the project was a very small communal one in which the water users themselves, rather than a separate irrigation agency, were to be responsible for its operation.

Obtaining the direct active input of water users is more difficult, and perhaps in some cases less feasible, in large government-initiated irrigation projects. Examples can easily be found where, in spite of the existence of financial autonomy, water users have little direct involvement in expenditure decisions (Box 10.5).

But financial autonomy gives the water users an important indirect role in these decisions. The fact that the irrigation agency must depend on the funds collected from the water users for its operating budget is apt to place certain constraints on expenditure decisions. Because of the sensitivity of the water users to the amounts that they are asked to pay for irrigation, governments and irrigation agencies are often reluctant to raise fees, or to allow the fees to exceed some maximum level. These considerations have the effect of restraining expenditures, thereby forcing the irrigation agency to make hard choices among alternative competing potential expenditures. In some cases, the level of irrigation fees is such a politically sensitive issue that the government may exert a high degree of control over expenditure decisions (Box 10.6).

10.5 *Setting irrigation fees: three practical issues*

To conclude this chapter on the setting of irrigation charges, we wish to examine three practical issues that often arise. The first of these is the issue of establishing project-specific irrigation fees, as opposed to fees that are uniform throughout an entire region or nation. The second issue (which typically would arise only in cases where project-specific irrigation rates are established) concerns the desirability of further differentiating rates according to differences in conditions within a given irrigation project. The third issue deals with the question of maintaining the real value of the irrigation fees over time in the face of inflation.

In the following discussion, we introduce each of these issues by posing the policy question that must be answered in any given situation. By doing so, we do not mean to imply, however, that a single 'correct' answer exists for all situations. Rather, our intent is to examine the types of

---

BOX 10.5

Role of water users in expenditure decisions under financial autonomy: examples from the Philippines and Korea

In the Philippines, the National Irrigation Administration (NIA) is a centralised, financially autonomous government corporation responsible for the operation of irrigation systems throughout the nation. Administratively, it is organised into 12 regions. Within each region are numerous projects for which the NIA is responsible. Farmers have virtually no representation or voice in the decisions that are made by the regional offices or at the individual project level. Only in situations where the NIA has turned partial responsibility for operation and/or maintenance to local water users' organisations do the farmers have any real direct voice in expenditure decisions. And given the centralised structure of the NIA, farmers would have little to gain by being involved in these decisions. Fees for a given project are not related to the magnitude of expenditures in that project, but rather are based on a national fee structure.

The situation in the Korean Farmland Improvements Associations (FLIAs) is different from the NIA in the Philippines because the FLIAs are decentralised, financially autonomous organisations. Thus the level of fees charged in a given association is quite closely related to the costs incurred in operating that association. In spite of this, the farmers have very little voice in the expenditure decisions. Instead, the government exercises close control over the expenditures of the FLIAs through mechanisms such as detailed regulations regarding the numbers and types of personnel that can be hired (see Box 10.6).

---

factors that need to be considered in arriving at a policy decision in a given situation.

10.5.1 *Should fees be project specific or uniform across projects?*

In some countries, irrigation fees are set at the national level, and water users in projects throughout the country are charged according to the same fee schedule. In other countries, fees vary from project to project. Still other countries have systems of fee schedules that fall between these two extremes – that is, projects may be categorised into a small number of groups, with a different fee structure for each group.

BOX 10.6

The role of the Korean Government in expenditure decisions of the financially autonomous irrigation associations

The Ministry of Agriculture and Fisheries exerts considerable control over the expenditures of the Farmland Improvement Associations (FLIAs). For example, the Ministry has developed guidelines for the staffing pattern of individual associations. These guidelines specify the number of departments, divisions, and staff members that an association is allowed to have. These numbers vary according to the size of the service area of the association. A large association (25 000 to 35 000 ha) is permitted four divisions and 13 sections, while the guidelines call for small FLIAs (5000 to 8000 ha) to have only one division and five sections. For the 73% of the FLIAs with service areas less than 5000 ha, only two or three sections, such as general affairs, finance and maintenance, are permitted. Since there is a specific limit on the number of staff in each division, the number of divisions determines the number of staff members permitted in an association. In this fashion, the Ministry attempts to place some limits on the operational expenditures of the associations.

The annual budgets of the FLIAs must also be approved by either the central or the provincial government. Detailed guidelines are laid down for the preparation of these budgets. For example, the budget guidelines for the cost of office heating specify the maximum temperature to which an FLIA office may be heated in the winter. The financial affairs of the associations are also supervised by the government through its financial audits of the associations, and through the requirements of its approval process for government loans to the associations.

Furthermore, the Ministry of Agriculture and Fisheries has established ceilings on the annual fees that can be charged by the FLIAs. These ceilings are formulated in terms of paddy, and translated into cash at the official government purchase price of rice. Reflecting the fact that the fees of all FLIAs have distinct components for operation and maintenance (O&M) and for repayment of capital costs, separate ceilings have been set for each component. For the O&M component, the ceilings established are 250 kg paddy per hectare for areas irrigated from reservoirs; 300 kg per ha for areas served by pumping stations; and 350 kg per ha for areas served by pumping and drainage stations. This generally acts as a limit on spending for O&M by the irrigation associations. The ceiling of the component for capital repayment has been fixed at 200 kg of paddy per ha since 1983.

There may also be a small number of special projects with individual fee structures. Examples of different situations are presented in Box 10.7.

BOX 10.7

Examples of irrigation fee structures

In Pakistan, irrigation fees are established by each province. Within a province, a fee structure is established for each of several categories of projects (or, as they are called in Pakistan, canals). Within each category of canals, the fee structure provides for differences in the fees per hectare depending on the crop grown (cotton, rice, sugarcane, maize, wheat, oilseed, vegetables) and on whether or not the area is served by deep tubewells designed to help control salinity. For example, in the Punjab province there are five categories of canals with respect to fees for vegetables, with the fees ranging from 34.4 to 42.4 Rupees per hectare.

In the Philippines, where rice is the predominant irrigated crop, the fees are denominated in paddy, although generally collected in cash. Throughout the country, farmers served by gravity irrigation systems are charged a uniform fee of 100 kg of paddy per hectare for irrigation during the wet season, and 150 kg for irrigation during the dry season. The one exception is in the Upper Pampanga River Integrated Irrigation System, which is the largest project in the Philippines. In this project, which was funded by the World Bank, and which presumably has better irrigation facilities than most projects, farmers are charged an additional 25 kg of paddy each season. In pump projects, individual projects have individual rates, ranging from 150 to 400 kg paddy during the wet season, and from 250 to 600 kg during the dry season. All of the rates noted above apply to the usual case of irrigated rice. For upland crops grown during the dry season, the rates are reduced by 40%.

In Korea, each FLIA establishes its own individual rate structure in order to meet its financial obligations. But as discussed in Box 10.6, the government does establish ceilings on the rates that can be charged by any FLIA.

In Nepal, irrigation fees in projects operated by the Department of Irrigation, Hydrology and Meteorology are nominally set at a fixed amount per hectare per irrigated crop, using two basic rates: 60 Rupees per hectare per crop and 100 Rupees per hectare per crop. With a few exceptions, the fee of 100 Rupees applies to projects that operate under separate Project Boards, while the 60 Rupee fee applies in other projects.

But this fairly simple rate structure has not been implemented consistently in all projects. In some cases the farmers are required to pay for only two crops, even if they grew a third irrigated crop in the winter. In at least one case, farmers are charged a flat amount of 200 Rupees per hectare per year irrespective of the number of crops grown. And some projects operate for many years with no assessment of fees.

Sri Lanka has recently attempted to impose an irrigation fee on land served by irrigation projects operated by the government. As originally designed, the fee structure was a very simple one whereby all land served by these irrigation projects was assessed at a flat rate of 250 Rupees per hectare per year.

To some extent, the question of the nature of the fee structure depends on the financial and organisational structure that exists for the delivery of irrigation services. In a situation characterised by decentralised financially autonomous irrigation associations, as is found in South Korea, it must be possible for fees to be set independently by each association. A single national structure for fees would not be compatible with the autonomy of the individual associations.

If a centralised financially autonomous irrigation agency exists, as is the case in the Philippines, fees could be either project specific or uniform. Fees in the Philippines tend to be uniform throughout the country, except in the case of pump projects, where the irrigation agency encounters particularly high operating costs. For these projects, the agency has resorted to establishing the fee structure on a project-by-project basis, presumably taking into consideration differences in the magnitude of the operating costs.

In the absence of financial autonomy, fees could be either project specific or uniform; however, with no inherent reason for project-specific rates, the administrative simplicity of relatively uniform rates is likely to be attractive. This type of rate structure prevails in Nepal and Pakistan.

It is sometimes alleged that uniform fees are better than project-specific fees because they are 'fairer'. As usual, however, this depends on one's perspectives about what is fair. Consider, as an example, two irrigation projects providing roughly the same quality of irrigation services in terms of quantity, timing and reliability of water deliveries. Both projects are served by the same reservoir; however, one project lies downstream from the reservoir and is served by a system of gravity canals, while the other lies above the reservoir, and is served by a system of canals fed by a large pump that pumps water from the reservoir to the canal network.

If one thinks of the irrigation services received by the farmers only in terms of water as an input for crop production, then farmers in the two projects receive the same service. From this perspective, charging them equal amounts seems fair. One could argue, in fact, that this is an example of providing horizontal equity – equal treatment of equals.

From another perspective, it is also reasonable to consider that these farmers are not equals. In terms of location with respect to the water supply, the force of gravity makes the farms (and thus those who farm them) unequal. More resources must be expended to provide water to the farms at the high elevation. From this perspective, those farmers receive an irrigation service that is different (and inherently more costly) from that provided to farmers at the lower elevation. This line of reasoning leads to the conclusion that charging the two groups of farmers the same amount would be inequitable.

The above example dealt with inherent differences in the O&M cost of providing irrigation to different groups of farmers. But what about situations where the primary differences are associated with the capital costs? If the water users are expected to repay a portion of the capital costs of irrigation, usually by some type of long-term loan arrangement, then project-specific charges may result in fees that are much higher for relatively new projects than for those that were built much earlier. While part of these differences may be due to differences in the 'real' cost of the irrigation facilities, a large portion may be attributable to the effects of inflation. Why should some farmers be penalised because of inflation, while others are not?

Although this problem may be perceived as one involving an unreasonably high fee for farmers in new projects, a careful analysis of the economics of inflation implies that the problem lies more in the large hidden subsidy being given to farmers in the older projects. This is illustrated with a simple example in Box 10.8.

One possible way to deal with this situation would be to maintain project-specific irrigation fees, but with loan repayments linked to a measure of inflation, using some mechanism such as adjustable interest rates. By eliminating most of the hidden subsidy created by inflation, this approach would allow for the use of project-specific irrigation fees that did not create such great inequities among projects. This would seem to be particularly appropriate in situations where an institutional structure

---

BOX 10.8
Inflation and the financing of capital costs: an illustrative example

Assume that in 1978 an irrigation project costing one million dollars was constructed under a policy that required the users to repay the capital costs over 25 years at no interest. Thus from 1979 to the year 2003 the annual payment to cover the capital costs is $40 000. In the absence of inflation, the only subsidy that exists is that resulting from the government's policy of charging a zero rate of interest, as opposed to a market rate.

If policies remain unchanged, a similar project built in 1988 would, in the absence of inflation, also cost one million dollars and result in annual payments of $40 000. The farmers served by the two projects would be paying the same amount for the cost of irrigation.

But now assume that ever since 1978, inflation has persistently prevailed at a rate of 15% per year. With this level of inflation, prices would approximately quadruple over a period of ten years. Thus the capital cost of the irrigation project built in 1988 would be approximately four million dollars, rather than one million. With unchanged policies, farmers served by the new project would be required to pay $160 000 per year, while their neighbours in the older project would be paying only $40 000. Obviously, the latter farmers are getting a better deal, and that better deal is the hidden subsidy caused by inflation.

To expose the hidden subsidy to scrutiny, we have prepared Table 10.2. For each of the years indicated, the table shows the 'nominal' amount paid. This is the actual 'out-of-pocket' payment – $40 000 current dollars each year. The next column shows the effect of inflation on the 'real' value of an 'out-of-pocket' dollar in 1978. Thus, for example, in 1979, one dollar had a 'real' value equivalent to only $0.8696 of a 1978 out-of-pocket dollar. In other words, one dollar in 1979 would purchase the same amount that approximately 87 cents would have purchased in 1978.

By multiplying these two columns together, we can find the value, in terms of 1978 dollars, of the actual payment. These amounts are given in the next column of the table. As shown in the next-to-last column, the actual out-of-pocket payment each year is 4% of the original capital cost. But in terms of the real values of the payment, expressed in terms of 1978 dollars, the percentage is less than 4% in the first year, and continues to drop further each year. These figures are shown in the last column of the table.

Table 10.2. *Illustration of the hidden subsidy of inflation*

Year	Actual payment (out-of-pocket dollars)	Value of $1 in the indicated year (measured in 1978 dollars)	'Real' value of payment (in 1978 dollars)	Payment as % of capital costs	
				Nominal value	Real value
1979	40 000	0.8696	34 800	4.0	3.5
1980	40 000	0.7561	30 200	4.0	3.0
1981	40 000	0.6570	26 300	4.0	2.6
1982	40 000	0.5718	22 900	4.0	2.3
1983	40 000	0.4972	19 900	4.0	2.0
1988	40 000	0.2472	9 900	4.0	1.0
1989	40 000	0.2149	8 600	4.0	0.9
1993	40 000	0.1229	4 900	4.0	0.5
1998	40 000	0.0611	2 400	4.0	0.2
2003	40 000	0.0304	1 200	4.0	0.1

In the absence of inflation, the government's policy regarding the repayment of the capital costs requires that 4% of the total cost be repaid each year for 25 years, giving a total cost recovery (at zero interest) of 100%. But as shown in the last column of the table, with an inflation rate of 15%, the real value of the total cost that is repaid in the first year is only 3.5%, and by the tenth year it is only 1.0%. If we were to calculate these values for each of the 25 years, and add them up, they would sum to give a total cost recovery (still calculated at zero interest) of only 25.9%.

Clearly, inflation has given the farmers served by the project built in 1978 a large subsidy that was not part of the explicit policy regarding cost recovery. If inflation continues at the same rate, the farmers in the project built in 1988 will also eventually receive the same subsidy. The difficulty, of course, comes because the farmers in the latter project are likely to compare themselves with the farmers in the other project at a particular point in time, such as in 1989. And when they do that, the situation appears quite inequitable. As can be seen from the table, in that year the $40 000 paid by farmers in the old project is equivalent in real value to less than 1% of the construction costs, while the farmers in the new project will be paying $160 000, which is equivalent in real value to about 3.5% of the construction costs.

of decentralised financial autonomy demands project-specific irrigation fees.

Alternatively, if it is deemed either undesirable or impractical to attempt to eliminate inflation's hidden subsidy, the establishment of a uniform fee schedule would ensure that the subsidy is made available to water users in all projects. Inevitably, however, this method will cause farmers in some projects to receive greater subsidies than farmers in other projects because a uniform fee structure does not account for real differences in the cost of irrigation.

10.5.2 *Should rates differ among water users within a single irrigation project?*

If a nation has established a system of fees that either is uniform throughout the country, or is uniform throughout a few major categories of projects, then the question of charging different rates among different water users within a single project does not arise. But the question does arise within the context of a structure of fees that is project specific.

The logic of project-specific fees is that water users in one project should pay for the costs of providing water to that project, and not some average cost of providing water throughout the country, or for a given category of projects. But this logic can also be extended to differences among individual water users within a project. If the cost of serving some users is greater than the cost of providing irrigation to others, then this economic logic dictates that higher fees should be charged to those for whom the cost of providing irrigation is greater.

This is done in some places, such as Korea (Box 10.9). But it must be kept in mind that as the number of distinctions increases, the administrative costs also increase. A balance needs to be struck between the differentiation of the fees in accordance with differences in costs, and the administrative ease of assessment and collection under a less differentiated fee structure.

The Paju FLIA in Korea discussed in Box 10.9 is a case in point. As noted in the Box, O&M fees are uniform within each of the five districts comprising the project. But this is a relatively new development. Prior to 1984, O&M charges within each district were differentiated according to some 20 categories of land. When asked about the reasons for the change to uniform O&M fees within each district, the officials of the Paju FLIA indicated that the purpose was to reduce the administrative burdens and costs associated with the previous complex structure of irrigation fees.

BOX 10.9
Differences in irrigation fees within a project: an example from Korea

The Paju FLIA can be used to illustrate the general type of structure that is found in the Korean FLIAs. Paju's irrigation fee consists of two components: one for O&M, and one for the repayment of the construction costs of the project. The O&M component of the water charge varies among the five districts, or subprojects, which together comprise the Paju FLIA. Within each district the O&M component is uniform for all the land. In calculating the O&M component of the water charge, a distinction is made between administrative costs and the direct cost of irrigation (pumping, operation of reservoir and canal gates, etc.) A single average per hectare cost of administration is calculated and applied throughout the entire area served by Paju, while the direct costs of irrigation are calculated separately for each of Paju's five districts.

With respect to the component of the water charge for the repayment of the project construction costs, four grades of land are recognised, based on the presumed benefits received as a result of the irrigation project. The highest charge is levied on land that becomes irrigated as a result of the project, and on which land consolidation has taken place. Newly irrigated land not yet consolidated is charged a lower amount. Land that already had some irrigation facilities prior to the construction of the irrigation project, and that has subsequently been consolidated, is charged a still lower amount, while the lowest charge is levied against previously irrigated land that has not been consolidated.

10.5.3 *How can the real value of irrigation fees be maintained in the face of inflation?*

Let us assume that an irrigation agency establishes a structure of fees that, at the time it is established, is at a level deemed reasonable in relation both to the agency's responsibilities for operating and maintaining the system and to the amount that water users can afford to pay. If, as is frequently the case, the rate of inflation is quite high, the real value of the fees established under this fee structure will soon become seriously eroded, so that the amount of funds collected will no longer be reasonable in relation either to the agency's needs or to the amounts that the users can afford. At the same time, political difficulties associated with an increase in the rate structure may make it extremely difficult to raise the nominal level of fees to restore their real value.

Tying the nominal irrigation fees to some general price index is one possible way to deal with this problem; however, this might be difficult to explain to the users, and therefore has the potential to create a considerable amount of dissatisfaction.

In situations where much of the irrigated area is devoted to one important crop (such as rice or wheat in many irrigation projects in Asia) an alternative is to establish the level of fees in terms of the quantity of this crop, rather than in monetary terms.

The fact that the fees are denominated in kind does not mean that they are necessarily collected in kind. Water users could still be required to pay the fees in cash, but the amount of cash necessary to satisfy the fee would depend on the price of the crop. The mechanism for establishing the price to be used each year would have to be specified. This approach has been used in the Philippines since 1975 (Box 10.10).

The advantage of establishing the level of fees in kind is particularly pronounced in the case of a centralised irrigation agency. Because of its national visibility, such an agency may encounter considerable political difficulty in raising rates, resulting in a decline in the real value of fees in the face of inflation. This is again illustrated by the Philippine case, where the real value of the irrigation rates declined substantially between 1946 and 1975, in part because any proposal to raise fees required the approval of the President, and thus became highly visible and subject to strong political pressures.

In situations where decentralised autonomous irrigation agencies exist, the problems of raising fees may be less severe because the decisions of individual irrigation agencies to raise rates have only local, rather than national, impact. They are therefore less likely to have the same degree of political visibility that a single national decision affecting all irrigation users could have.

10.6 *Summary*

Financing irrigation with user fees requires consideration of both the costs of providing irrigation and the benefits that the users receive from irrigation. In many cases, fees designed to recover the total costs of irrigation would be so high that many farmers would be left with a lower income than they would have had in the absence of irrigation. In practice, however, farmers are seldom asked to pay fees that cover the total costs of irrigation. Rather, the emphasis is more likely to be on the costs of operation and maintenance. This focus on O&M costs is often appropriate, since it emphasises obtaining the funds needed to provide irrigation services to the farmers on a continuing basis, while ignoring the capital costs which, because they have become sunk costs, are irrelevant to future operating decisions.

In well-functioning irrigation systems, the water users should be able to pay fees that are high enough to cover the normal costs of operation and

---

BOX 10.10

Setting irrigation rates in kind: experience from the Philippines

Irrigation fees have been levied in the Philippines since at least 1946. Originally the rates were established in monetary terms. But over time, inflation severely eroded the real value of these fees, which, for political reasons, were difficult to raise. Since 1975, the irrigation fees paid by farmers have been denominated in terms of paddy. The farmers may pay either in kind or in an equivalent amount of cash. The cash value of the fee is based on the government's support price of paddy, and therefore increases with any increase in the support price.

Table 10.3 shows the dry season irrigation fee rates for gravity irrigation projects from 1946 to 1984. The nominal and real (in constant 1984 Pesos)

Table 10.3. *Nominal and real irrigation fee rates for the dry season in Philippine gravity irrigation systems, 1946–1984*

Year	Irrigation fee (kg paddy per hectare)	Nominal value of fee (current Pesos per hectare)	Real value of fee (Constant 1984 pesos per hectare)
1946–66	NA	12	–
1966–75	NA	35	–
1975	150	150	514
1976	150	165	516
1977	150	165	471
1978	150	165	439
1979	150	195	449
1980	150	210	420
1981	150	226	411
1982	150	248	414
1983	150	267	399
1984	150	335	335

values of the cash equivalent of these rates are also shown. It is clear that this mechanism has provided a degree of indexation against inflation. Between 1975 and 1984, the real value of the Peso declined by about 71%. But even though the irrigation fee rates have not been raised since 1975, the decline in their real value over this same time period has been only approximately 35%.

Thus, although fees fixed in kind do not guarantee that their real value will remain constant, and although there may be short periods of particular difficulty when nominal commodity prices drop in spite of a general inflationary trend in the economy, such a system of fees is likely to be a considerable improvement over fees fixed in monetary terms.

---

maintenance. Users who are asked to pay for the costs of operation and maintenance have a legitimate interest in the decisions about operation and maintenance expenditures, and may desire some role in the process to see that these expenditures are kept within reasonable bounds.

If responsibility for the operation of a nation's irrigation project lies with decentralised financially autonomous irrigation agencies, then it will be necessary to have different levels of fees in each project. Within an individual project, fees might be further differentiated. To the extent that differences in fees reflect differences in the inherent costs of irrigation, this may be seen as both equitable and efficient. But to the extent that the fee differentials reflect arbitrary differences created by the implicit subsidies associated with inflation, they may be seen as very unfair.

In situations where most of the irrigated area is devoted to the production of a single crop, denomination of the irrigation fee in kind rather than in cash may help protect the irrigation agency from having inflation erode the real value of the funds it collects from the water users. This can be particularly important in the case of a centralised but financially autonomous irrigation agency, where the agency's national visibility may make it politically difficult for it to raise the fees to offset the effects of inflation.

11

Collecting irrigation fees: fostering a willingness to pay

11.1 *Introduction*

Even the best designed system of irrigation fees can be destroyed quickly by low rates of collection of the fees assessed. Low collection rates create severe inequities. It is obviously inequitable for some water users to pay for their irrigation services while others do not. And if the irrigation agency is financially autonomous, this inequity is even greater because the failure to collect fees from some water users must be reflected in higher rates of assessments. Thus those who pay their fees are in effect paying not only for the services they receive, but also for the services received by those water users who have failed to pay. If such a situation prevails for long, the credibility of the fee system is likely to deteriorate, with rates of collection dropping even further.

Discussions about fee collections sometimes focus exclusively on questions of enforcement. Although enforcement is important, we wish to emphasise that it is only one aspect of the collection of irrigation fees. More generally, we need to be concerned with all the conditions (including enforcement) which create a willingness to pay on the part of the water users. This means that we need to be concerned with positive incentives for the water users to pay, as well as with penalties established as a means of enforcement. Furthermore, we need to consider how institutional conditions can affect the willingness of the users to pay their irrigation fees. Finally, we must remember that rates of collection depend not only on the willingness of the users to pay, but also on the 'willingness' of the collecting agency to collect. In other words, the collecting agency needs incentives to encourage it to allocate an appropriate level of resources toward the collection of irrigation fees. This will allow the agency to take the initiative with positive efforts to collect the irrigation fees, rather than simply to wait passively for users to make their payments.

In this chapter we first consider some of the institutional factors that are likely to affect both the willingness of users to pay and the willingness of the agency to collect. Then we examine various strategies of collection, including types of positive incentives to encourage payment, types of penalties that can be imposed on those who do not pay, and other conditions that can encourage higher rates of payment.

11.2 *Institutional factors*

Several institutional factors can affect the rates of collection of irrigation fees. In the following discussion, we have grouped these factors according to whether they impinge primarily on the water users or primarily on the agency responsible for fee collection.

11.2.1 *Factors related to the water users*

Ownership of facilities

Perceptions of the water users regarding the ownership of the irrigation facilities may affect their willingness to pay irrigation fees. Water users often appear more willing to pay fees when they perceive themselves or their community as the owner of the irrigation facilities than when they perceive the government to be the owner. Many water users may be reluctant to pay fees for government-owned facilities on the grounds that it is the government's responsibility to operate and maintain the facilities (Box 11.1).

Disposition of the fees

Farmers who are being asked to pay irrigation fees are likely to be interested in what happens to the fees that they pay. Two general approaches to the disposition of the fees can be identified. One, typical for situations of central financing, is for the funds to flow through the collecting agency directly into the government treasury and to become part of the government's general revenues. Little or no linkage exists between the amount of funds collected from the water users and the amount of funds made available to provide services for them. As a result this approach creates no inherent incentives for the water users to pay the fees.

The alternative approach, which prevails with financially autonomous irrigation agencies, is for the fees collected by the agency to be retained

---

BOX 11.1

Unintended effects of government assistance to farmer-managed irrigation projects

Many countries have irrigation systems that have been constructed and operated without any government assistance. Often these 'farmer-managed' irrigation systems utilise technology that has a low capital cost but a high labour cost. For example, the headworks often consist of brush or rock diversion weirs that must be replaced each year after they are washed out during the season of heavy rains. Such systems are clearly owned by the local communities that built and operate them, and these communities have been able to mobilise the resources (in the form of labour, materials or cash) needed to continue to operate and maintain them. In other words, these communities have been able to effectively impose and collect irrigation fees for the operation and repair of these community-owned facilities.

In recent years, government agencies in several countries have attempted to improve such irrigation systems, often by replacing the temporary facilities by 'permanent' ones built with concrete. All too frequently, however, this intervention by the government reduces the ability of the local community to mobilise the resources needed for the continued operation and maintenance of the facilities. In part, the problem seems to be that the water users no longer regard the facilities as their own. The new 'improved' facilities belong to the government, and it is thus the government's responsibility to operate and maintain them.

---

by it for use in the operation and maintenance of the irrigation facilities. Higher rates of collection mean more funds available for O&M, a fact that creates a potential incentive for the users to pay their fees.

Financially autonomous irrigation agencies can exist along a continuum with respect to centralisation. At one extreme, a single irrigation agency is responsible for the operation of irrigation systems throughout an entire nation. At the other extreme, each individual irrigation project is operated by its own financially autonomous organisation. The greater the degree of financial decentralisation (and thus the more direct the relationship between an individual farmer's payments and the operation and maintenance of the facilities serving that farmer) the more effective the incentive for payment created by financial autonomy is likely to be.

Farmer-managed irrigation systems (see Box 11.1) represent the extreme case of decentralised financial autonomy, and incentives for payment of the fees are very strong. The irrigation associations of Taiwan and South Korea represent a somewhat less extreme case of decentralised financial autonomy. In the case of Korea, a total of 103 financially autonomous Farmland Improvement Associations (FLIAs) are responsible for the operation of all but the smallest-scale irrigation facilities throughout the country. This indicates a considerable degree of decentralisation; however, in the past as many as 699 associations have been responsible for the operation of these facilities. The sharp reduction in the number of associations thus represented a significant increase in centralisation. Within an individual FLIA, however, provision is sometimes made for distinctions in the fees between areas irrigated from different water sources. This increases the link between each farmer's payments and the cost of providing irrigation to that farmer.

The National Irrigation Administration (NIA) of the Philippines is a good example of the extreme case of centralisation in a financially autonomous irrigation agency. It is responsible for the O&M of government irrigation projects throughout all 12 administrative regions into which the country is divided. All irrigation fees collected from the farmers are retained by the NIA, but are not necessarily used for the operation and maintenance of the specific project from which they were derived. In general, the irrigation fee rates are set at the national level, so that farmers in gravity irrigation projects throughout the nation pay the same level of fee. The effect of this arrangement is for funds collected from projects whose O&M costs are below average to be used to subsidise the O&M costs of projects whose costs are above average. The lack of a direct relationship between the project from which fees are collected and the funds available for O&M in that project probably weakens the incentive for individual farmers to pay their fees. This may be one of the reasons why, in spite of the NIA's very substantial efforts at fee collection, rates of collection of irrigation fees in the Philippines remain less than satisfactory.

Accountability of irrigation agency personnel to water users

It seems reasonable to expect that, other things being equal, water users will be more inclined to pay their irrigation fees if they perceive that their concerns and preferences are taken into account when operating decisions are made by the personnel of the irrigation agency. This is more likely to be the case when the personnel of the irrigation agency are accountable to the water users. Irrigation systems in Taiwan, where collection rates are generally very high, provide a good example (Box 11.2).

BOX 11.2

Accountability of irrigation managers in Taiwan

In an article entitled 'Irrigation systems in Taiwan: management of decentralized public enterprise', Martin Abel has discussed the ways in which managers of irrigation facilities are accountable to the farmers. He notes:

> **A distinctive feature of the irrigation of Taiwan is that the systems are essentially owned and managed by the farmer-users of the water. Thus the managers of the irrigation systems work for the farmers. The irrigation associations, which are farmer cooperatives, can hire or fire managers depending upon their performance. Even when some members of management are appointed by the government, they are expected to be responsive to the needs and desires of the members of the irrigation association.**
>
> **. . . (T)he rewards to management are determined by the elected representatives of the members of irrigation associations. And there is evidence that irrigation associations do reward good management and do penalize poor management. The reward structure includes financial returns to management, promotions, and nonmonetary recognitions such as prizes.[1]**

This accountability is also linked to the fact that the irrigation associations are financially autonomous. Again we quote Abel:

> **Another important aspect of the incentive system is the interrelationship between the collection of irrigation fees and financing the operations of an irrigation association. The operating budget of an irrigation association depends directly on the collection of water fees from farmers. In order to preserve their jobs, the technical and administrative staffs of an irrigation association have a strong interest in ensuring the collection of fees. If collections are poor, revenue will not be adequate to cover operating costs and will eventually result in a reduction in the size of the staff of the association.**
>
> **The willingness of farmers to pay their fees depends heavily on how well the irrigation associations are operated, i.e. the amount and timeliness of water received. The better the system is**

managed, the more willing the farmers will be to pay their fees. This is also true for voluntary farmer participation in certain operations of the system, such as controlling the release of water into fields, performing maintenance work on the portion of the system located near their farms, etc. Thus job security (and) levels of remuneration for management personnel are tied directly to how well a system is managed.[2]

Collection commissions

Some irrigation agencies find it too costly for them to attempt to collect fees directly from individual farmers. Instead, they prefer to assign responsibility for collection to an intermediary to whom the irrigation agency pays a commission in the form of a certain percentage of the fees collected. In some situations the intermediary with whom the irrigation agency contracts is a water users' association. When this occurs, an additional incentive is created for individual water users to pay their fees, since the water users as a group can retain a portion of the fees paid. Furthermore, this mechanism gives the entire group of farmers a financial interest in each farmer's payment. Group pressures are likely to be brought to bear on those who have failed to pay. These pressures, both actual and potential, further increase the incentives that farmers face to pay their fees.

The National Irrigation Administration of the Philippines has experimented with a variety of such approaches to collect fees in some of the government irrigation projects. As discussed in greater detail in Box 11.3, these arrangements have frequently involved a commission that has been structured so that the percentage of the fees collected that can be retained by the water users' associations actually increases with the rate of collection. Cases have been reported of associations that have prepaid the fees of some members who have not yet paid, in order to obtain the benefit of the higher percentage commission. The association subsequently takes responsibility for getting their delinquent members to pay. Because the association is composed of farmers who know and interact regularly with each other, it is likely to be able to exert much more effective social pressure on the delinquent members to pay than could the NIA. In economic terms, the farmers' knowledge and opportunity for direct interaction is a valuable resource of the water users' association that the NIA lacks. It is the possession of this resource that makes it possible for the association to collect the fees at a lower cost than the NIA would have to incur. Thus the commission arrangement creates financial benefits both to the NIA and to the water users' association.

BOX 11.3

Providing group-based financial incentives for paying irrigation fees: examples from the Philippines

In the Philippines, the NIA has tried a number of approaches to giving incentives to water users' organisations that would result in higher effective rates of fee collection. In these approaches, the NIA has generally tried to combine the delegation of certain O&M responsibilities to the water users' organisations with the provision of incentives to encourage these organisations to take an active role in fee collection.

For example, under the so-called 'lateral turnover' arrangement, the water users' association contracts with the NIA to undertake maintenance of the lateral canal, with the association to be paid at a specified rate per kilometre of canal. To the extent that the work can be done at a lower cash cost (by encouraging farmers to contribute unpaid labour) the association is able to earn a cash income. Furthermore, the association is allowed to retain 2.5% of the fees it collects from its members if it achieves a target rate of 70% collection. If the collection rate rises to 100%, the association can retain 3% of the collections.

Under another arrangement, the water users' association is given full responsibility for system maintenance, but no cash payment. Rather, the association is allowed to retain a much higher portion of the amount of irrigation service fees it collects from its members. For collections below 50% of the aggregate amount due, the association is allowed to retain 35% of the funds collected. For all collections above this 50% figure, the association is allowed to retain 65% of the amounts collected.

11.2.2 *Factors related to the collection agency*

Unless the agency responsible for collecting irrigation fees undertakes a strong and active collection effort, rates of collection are likely to be unsatisfactory. But to undertake this effort requires that the agency commit both human and financial resources to the collection process. It is unlikely to do so unless appropriate incentives are provided.

Financial autonomy gives an irrigation agency a strong incentive to devote resources to the collection of irrigation fees. In this sense, financial autonomy is comparable to a 100% commission on fees collected. Not only does financial autonomy give an irrigation agency financial incentives to devote resources to the collection process, but it gives the agency the incentive to devote the optimum amount of resources to this process. As with other economic activities, diminishing returns can be expected to the expenditure of resources on the collection of fees. We should therefore not expect that an autonomous agency would necessarily attempt to maximise the rate of collection. Rather, the agency would attempt to balance additional expenditures with their likely effect, both in the short and long run, on revenues from fee collections.

The contrast with central financing is very clear. From the perspective of a centrally financed irrigation agency, the economic optimum expenditure on fee collection would be zero unless commissions are provided. The reason is simple: regardless of the amount of fees collected, no additional funds will be available to the agency as a result of its expenditures on fee collection. Bureaucratic directives may cause a centrally financed irrigation agency to undertake some fee collection activities, but it is likely either that compliance with the regulations will be minimal, or that efforts will be made to change the situation. The examples from the Philippines and Nepal discussed in Box 11.4 illustrate these points well.

BOX 11.4

Incentives to collect fees: examples from the Philippines and Nepal

The case of the National Irrigation Administration (NIA) of the Philippines

As discussed in some detail in Box 9.6, under its original charter the NIA operated primarily as a centrally financed government agency. Although it was responsible for collecting fees from the water users, the resulting revenues had to be turned over to the government treasury. Writing about conditions during this period of time, Benjamin Bagadion, a former Assistant Administrator of the NIA notes that 'among NIA systems personnel there was no sense of urgency to increase collections since the amount appropriated by the central government did not depend on the amount collected from the water users'.[3]

The revision in the NIA's charter in 1974 that allowed it to keep all the fees that it collected brought about dramatic changes. The new financial arrangements gave the NIA a much greater incentive to collect fees than had previously existed, and led to a considerable increase in the importance that it attached to rates of fee collection. The NIA's efforts have concentrated on placing more internal emphasis on fee collection, and on providing a variety of incentives for payment.

The case of Nepal

Responsibility for the collection of irrigation fees in Nepal was once added to functions of the Land Revenue Office, which also collected the land tax. Apparently no financial incentives were given, and the Land Revenue Office lacked the staff needed to undertake the collection of irrigation revenues. When additional staff were not provided, the Office refused to continue to collect the irrigation fee. Responsibility for assessment and collection of the fees was then shifted to the centrally financed agencies responsible for the operation and maintenance of the projects.

But in the absence of financial autonomy, this responsibility represented only a financial burden to the irrigation agencies. As a result, it appears that little importance is attached to the collection of irrigation fees.

A variety of administrative problems commonly reported in the mid-1980s reflect the low priority given to fee collection. Difficulties were encountered in determining the land actually irrigated; ambiguities arose with respect to responsibility for payment in cases where the land was not operated by the landowner; farmers were expected to come to the project office to pay the service charges, even though no bills were sent directly to them; and no effective system of penalties for non-payment had been implemented, at least in areas served by surface water.

In recent years responsibility for the collection of irrigation fees was again transferred. Without the appropriate institutional arrangements, responsibility for collecting irrigation fees appears to be a responsibility that no one wants and that no one will undertake in a satisfactory fashion.

11.3 *Strategies of collection*

We hope that the previous section has demonstrated clearly the importance of institutional factors with respect to the collection of irrigation fees. But most of these institutional factors involve long-term considerations over which an irrigation agency has little control. In the shorter term, what can an agency do to maximise the net revenues generated from fee collection? In this section, we examine several types of action that could be part of an irrigation agency's collection strategy, namely: providing positive incentives for the water users to pay their fees; providing enforceable penalties for non-payment; creating other general conditions that enhance the likelihood of payment; and avoiding excessive collection costs.

11.3.1 *Providing positive incentives for water users to pay*

Irrigation agencies may provide positive financial incentives designed to encourage prompt payment of irrigation fees. Some of these incentives are targeted at the level of individual farmers, while others focus on water users' groups. For example, for a number of years the NIA of the Philippines offered a 10% discount on the assessed fee to any farmer who paid on or before the due date. And, as discussed in Box 11.3, the NIA has also experimented with a number of approaches to providing incentives to water users' groups.

But incentives need not be limited to financial considerations. Incentives linked to social factors may also be helpful in encouraging payment of fees. In South Korea, the Farmland Improvement Associations (FLIAs) attempt to encourage prompt payment through competitions. Within the area served by each field station of the FLIA, prizes may be given to the first three villages to achieve full payment of the fees. Although the prizes are monetary, the amounts are modest. For example, in one relatively large FLIA, the first prize in 1985 was equivalent to approximately 80 kg of paddy (unhusked rice). The key motivating force of the award is the recognition and status that it carries.

Social incentives can also operate through personal linkages between the users and those responsible for operating the irrigation system. In large government irrigation projects in Indonesia, for example, operational and financial responsibility for the tertiary distribution system is decentralised to the village level. In some parts of the country, the village officials responsible for operating the facilities are reimbursed for their services by payments made to them by the farmers. These payments are not necessarily fixed charges or 'taxes'; rather, they may be what the Indonesians call 'feeling' (*pangrasa*) payments. The amount paid by each water user depends on his feelings regarding the quality of the services received, the outcome in terms of crop production, and the social relationships that exist between the farmer and the local officials managing these tertiary facilities. The incentives for payment are thus to be found in both the social and the agricultural linkages between the farmers and the local officials.

11.3.2 *Providing enforceable penalties for non-payment*

Most nations have formal penalties for water users who do not pay their irrigation fees. But all too often these penalties have little practical meaning because of enforcement problems. Enforcement may founder because of technical factors associated with the water distribution system, because of cumbersome administrative and legal procedures, or due to the lack of political will. Since a system of penalties that cannot be enforced may be worse than no penalties at all, the question of enforceability needs careful consideration in an agency's strategy for collecting irrigation fees.

Termination of water deliveries

Any farmer who finds water to be valuable is unlikely to risk its loss by failure to pay the required irrigation fee. Termination of water deliveries to those who refuse to pay their fees is, therefore, potentially one of the most effective penalties that could be imposed. Although some examples of this approach can be found (such as in some projects in Mexico and China where the operating agency will only deliver water to farmers who can produce receipts showing that prior payment has been made) they are the exception rather than the rule. Many irrigation agencies lack the degree of physical control over water distribution needed to make this approach work. Particularly in large projects serving many small farmers, it is often infeasible to prevent the distribution of water to an individual farmer whose neighbours continue to receive water. And some countries are unwilling even to consider this type of penalty. In South Korea, for example, where irrigation is primarily used for rice production, the importance of rice both to the nation and to the individual farmers is considered to be so great as to render unacceptable any proposal that could deny water to the farmer.

In cases where halting water deliveries is in principle politically acceptable, but where technical factors make it infeasible to implement at the level of the individual farmer, an irrigation agency may consider the possibility of withholding water from groups of farmers. This is most likely to be attempted in pump projects, where the irrigation agency is keenly aware of the marginal costs of supplying water. For example, in the Philippines the NIA has stated a policy of ceasing to operate any pump project in which the total amount collected from fees assessed for the previous season is below 90% of the amount due. But this policy has been easier to state than to implement. In several cases the NIA's efforts to implement the policy have been thwarted by external political pressures that the farmers were able to mobilise and bring to bear on the NIA.

A possible alternative to the complete termination of water deliveries to groups of farmers has been suggested for a World Bank funded project in India. Under this proposal, water deliveries would be organised and monitored to units comprising groups of farmers served by a defined service area. Delivery of the full water allotment to the service area would occur only if the total water bill for the area had been fully paid. If payment had not been made in full, water deliveries would be reduced in proportion to the percentage of the total bill that had not yet been paid.

Financial sanctions

Where termination or reduction of service is not feasible, other approaches to enforcement are needed. Many countries have regulations that provide for the imposition of fines for improper water use and for failure to make prompt payment of irrigation service fees. The effectiveness of such fines depends on several factors, including the extent to which they are actually implemented, and their magnitude relative both to the value of water and to interest rates. For example, fines levied in Mexico for illegal use of water are of questionable effectiveness because they are reportedly less than the value of the water taken.

Fines are one component of the enforcement system used in South Korea (Box 11.5). Farmer-managed systems in Nepal regularly impose fines on members for being absent when required to participate in maintenance work on the system. The organisations are very successful in collecting the full amount of fees and fines that are charged. The membership brings social and, sometimes, economic pressure to bear on members who refuse to pay. An example has been reported in one system where the members took the cooking utensils of a farmer who had refused to pay and threatened to sell them to obtain the amount of cash due. The action was successful: the reluctant farmer paid his fees, and all the members were made aware of the organisation's determination to collect the fees assessed.

Legal sanctions

Legal sanctions for failure to fulfil financial obligations are imposed in some countries. Formal provisions sometimes exist for foreclosure and sale of the land of delinquent water users (Box 11.6);

BOX 11.5

Enforcement of irrigation charges in South Korea

Under the Rural Modernisation Promotion Act of 1970, the Farmland Improvement Associations (FLIAs) of South Korea were granted the power to collect water charges under the general taxation authority given to local governments. The FLIAs are thus responsible for fee assessment, billing, collection and enforcement.

The FLIAs have established financial penalties for late payment of the water charges. These penalties were first introduced in 1952 in response to problems of late payment and non-payment of the fees. The current penalty for late payment is equivalent to 5% of the assessed fee if payment is made within the first month after it was due. For each of the five succeeding months, an additional 2% penalty is added. If the fee has still not been paid at the end of six months (when the total penalty has reached 15%), the FLIA can initiate legal proceedings to sell the farmers' assets (excluding farmland, which by law cannot be sold for non-payment of taxes) to recover the charge. The police then sequester assets of the farmer valued at the amount owed, and can sell them after 15 days if the farmer has still not paid. However, it appears that this procedure is very rarely implemented, as most small farmers have few assets that could be sold.

According to the Chairman of the Paju FLIA, interviewed in 1985, legal action had never been taken by the association against any farmer; however, a number of farmers had been penalised for late payment. Data for 1984 for the Paju FLIA revealed that a total of 3 304 700 *won* had been collected in penalties from 418 farmers (about 2% of the members of the FLIA) for late payment. The amount collected in penalties was less than 0.2% of the total amount of water charges collected by the FLIA in 1984.

---

however, in most countries such extreme actions are generally considered to be unacceptable. Even for less extreme penalties, the difficulties associated with taking court action against numerous small farmers frequently undermine the effectiveness of legal sanctions. In a few countries, however, such as Spain, simplified legal mechanisms have been established specifically to deal with irrigation problems.[4] This type of institutional change greatly enhances the practicality of using legal sanctions to enforce the rules surrounding irrigation, including those regarding the payment of fees.

BOX 11.6
Legal sanctions: examples from the United States and Nepal

In some irrigation projects in the United States, financially autonomous irrigation districts have the authority to levy water charges to meet operating costs, and to make required repayments of construction costs to the federal government. These charges are assessments against the value of the land, so that failure to pay results in a lien on the land. As a last resort, the district may foreclose and sell the land to obtain the delinquent funds. Furthermore, the United States federal government has sometimes required that the irrigation districts impose joint liability on their members for the repayment of the construction costs owed to the federal government. As a result, if a district falls into arrears on its payments to the government, no individual landowner within the district can obtain a clear title to his or her land.

In Nepal, irrigation fees are treated in the same fashion as land taxes. Foreclosure and sale of the land is not, however, practised as a method of enforcement. Rather, the law provides that a land owner wishing to complete a legal sale of land must first pay any outstanding irrigation charges.

Social sanctions

In some countries social pressures and sanctions may be an important method of encouraging water users to fulfil their financial obligations, as well as to obey water allocation rules. These sanctions may take a wide variety of forms. In some cases those who are delinquent in paying their fees may not be allowed to participate in various local social organisations. In one case observed during a research project in the Philippines, farmers who were cleaning the canals of a small farmer-managed system dumped a mound of earth, in the form of a grave, on the land of a farmer who had not shown up to participate in the work. In South Korea it has been reported that on occasion a farmer who has not paid his fees may return to his house after a day working in his fields to find that his neighbours have removed its roof. The common thread in all of these sanctions is that local people, acting through informal mechanisms, make their displeasure known to the delinquent party in ways that are obvious both to the individual and to others, thereby causing discomfort to the delinquent party.

Although enforceable sanctions are important to a viable system of water charges, if the sanctions and their enforcement become too strong, there is a danger that the accountability linkages between the irrigation agency and the water users may be weakened. With very strong sanctions, the irrigation agency no longer needs to be seriously concerned (at least in the short run) that water users might withhold payments as a means of protesting the quality of service (Box 11.7). What is needed is a balance between the emphasis placed on providing a satisfactory service for which the farmers will be willing to pay, and enforcement that ensures that the 'free rider' problem is kept under control.

BOX 11.7
Enforcement and accountability linkages in South Korean FLIAs

Robert Wade, who undertook an intensive study of one FLIA in South Korea, argues that in Korea because both the incentives for prompt payment and the coercive sanctions for non-payment are extremely strong, the accountability linkages between the agency personnel and the water users is significantly weakened.[5] In effect, the Korean farmers cannot consider the withholding of irrigation fees to be a viable way of expressing any discontent they may have with the quality of the irrigation service.

11.3.3 *Creating conditions favourable to payment*

In addition to establishing direct incentives and penalties to encourage payment of irrigation fees, an irrigation agency can improve the prospects for fee collection by creating a variety of general conditions that are conducive to collection.

Shifting responsibility for collection to intermediaries

Personnel of irrigation agencies often are too physically and socially removed from the individual water users for them to be effective collecting agents. Collections may be enhanced through the use of intermediaries (Box 11.8). In some cases the intermediary may be a water users' organisation, while in others it may be an individual who is given this responsibility. Although rare exceptions can be found, some type of payment is generally needed if the intermediaries are to be effective. When the financial incentives given to the intermediaries are strong enough, they may actually pay the fee of a delinquent water user in anticipation of being able to collect from the user at a later date.

These arrangements are likely to be cost-effective ways of increasing fee collections because they rely on people who have more intimate knowledge of the conditions and circumstances of the individuals who are supposed to pay, and who are thus in a better position to formulate and

---

BOX 11.8
Using intermediaries for collecting irrigation fees: an example from South Korea

In South Korea, two types of intermediaries are used to facilitate the payment of the irrigation fees. Each village served by an FLIA has a designated farmer representative (officially known as the *Huong Nong Gye* leader) who serves as a liaison between the FLIA and the water users in the village. One of the responsibilities of these individuals is to get farmers to pay their fees. The position carries no financial compensation, and it is reported that people generally do not wish to serve in this capacity. Village leaders frequently give this responsibility to older and prestigious people.

The actual payment of the fees involves a second intermediary, which is the county or subcounty cooperative unit of the National Agricultural Cooperative Federation. Every year the local FLIA signs an agreement with the county cooperative authorising it to receive, on behalf of the FLIA, the payments of the farmers' irrigation fees. The county cooperative then notifies its subcounty cooperatives of the agreement, and authorises them to receive the payments of farmers to be credited to the account of the FLIA. Individual farmers may then pay their bills either at the county cooperative or at any of its subcounty offices.

The cooperatives receive no direct financial payments in exchange for providing this service; however, they benefit indirectly through the additional business that is likely to be transacted by the farmers when they come to pay their irrigation fees.

This use of the cooperatives as intermediaries probably improves the administrative efficiency of fee collection by allowing for greater specialisation of functions. The cooperatives already have personnel specialised in undertaking financial transactions with the individual farmers; therefore, it is not necessary to establish a separate group of personnel responsible solely for the collection of irrigation fees.

---

time their collection efforts in ways that are likely to be successful. Both the knowledge that the intermediary possesses about individual water users, and the social relationships between the intermediary and the water users are valuable resources that can be used to improve fee collection.

Making direct contact with the water users

Collection rates may sometimes be improved if those responsible for collections make direct contact with the farmers. In the Philippines, it has been reported that farmers are less likely to refuse to pay their irrigation fees when staff of the irrigation agency contact them immediately after harvest. By making direct contact (particularly at a point in time when it is obvious that the farmer has the means to pay) the agency may place the farmer in a position of losing face by refusing to pay. Assigning personnel to go to the individual farmers to request payments is, of course, costly. But the effort may be worth the cost if collections increase substantially. One of the advantages of using intermediaries (such as the *Huong Nong Gye* leaders in Korea discussed in Box 11.8) is that it provides a mechanism that facilitates direct contact with water users at a lower cost to the irrigation agency – and probably often at a lower cost to society.

Providing for payment in kind

It has sometimes been found that payment rates can be increased if the irrigation agency is willing to accept payment in kind, rather than in cash. Thus, in the Philippines, the NIA for a number of years collected most of its fees in the form of paddy rice, rather than in cash. Collections in Korea were also in kind for a period of time.

Collections in kind have the advantage of making it more difficult for the farmer to refuse payment. If a collection agent arrives at a farmer's house in which a substantial amount of grain is stored, it is difficult for the farmer to claim that he or she has no means of paying the fee. On the other hand, collection in kind greatly increases the cost of collection. The irrigation agency must now find means of transporting, storing, and selling the grain that it has purchased. Often problems of quality arise. In the Philippines it was found that farmers often sold very wet grain to the NIA in payment for their fees. This increased the losses due to spoilage, and the costs of handling and drying the grain to prevent spoilage. As a result, an irrigation agency may resort to collecting fees in kind in situations where the farmer payment rates have been poor, but later switch back to cash collections when it appears that farmers have generally accepted the idea of paying their irrigation fees.

Linking fee payments to other payments

Some countries already have effective arrangements for collecting certain taxes or fees from the farmers. In these situations, collection of irrigation fees may be facilitated by administratively linking them to the existing arrangements. For example, during the early 1980s, the governments of Indonesia considered various proposals to generate increased revenues for operating and maintaining its irrigation systems. One proposal involved linking an irrigation fee with the land tax. Although revenues from the land tax had generally not been available for use in operating irrigation systems, the tax was a well-established and reasonably effective administrative mechanism for billing and collection from individual landowners. It was thus suggested that an irrigation fee could be implemented by raising the land tax on irrigated land, and earmarking the additional revenues for use in irrigation O&M. This would be administratively efficient, as it would eliminate the need for a separate administrative structure for collecting the irrigation fee. In addition, it had the advantage of avoiding a direct charge on irrigation water, which was seen as somewhat in conflict with religious and cultural traditions.

An alternative approach used in the Gezira irrigation project in the Sudan involves an automatic check-off arrangement for paying the irrigation fee. The role of the government in this project is comprehensive. It owns the land, sets the cropping patterns, provides the inputs, and markets the crops produced. As a result, the water users operate primarily as tenant farmers, receiving from the government at the end of the year the difference between the value of the crop they produced (based on the price that is established by the government) and the cost of the inputs that the government has provided. Under this type of an arrangement, it is easy for the government to include an irrigation fee in its calculations of the cost of the inputs it has provided. Since the government has an effective monopoly on the marketing of the crop, collection of the irrigation fee is virtually assured.

These arrangements linking irrigation fee payments to other payment mechanisms increase the ease and rates of collection; however, they also tend to eliminate any accountability linkages, which a system of irrigation fees might otherwise create, between those operating the irrigation facilities and the farmers. By making it difficult or impossible for the farmers to avoid paying the irrigation fee, the government effectively removes from the farmers the ability to withhold (or to threaten to withhold) payment for irrigation fees as a protest of inadequate service or excessive cost. It is even possible that the farmers will be completely unaware of the existence of an irrigation fee. In the case of the check-off, the amounts that the farmers receive for their crop depends on several factors (including the price of the crop) which, from the farmers' perspective, are all arbitrarily controlled by the government. The situation with the land tax is similar.

Thus, while linking irrigation fees to other payments may increase the ease of collection, most or all of the benefit of establishing the irrigation fees may be lost! If the revenues generated by the fees are specifically earmarked for irrigation O&M, then the system of fees retains some value. But if not, it will not improve the funding of the irrigation agency; it will not build accountability linkages between the agency and the water users; and it will not encourage the farmers to be more efficient in their use of water. And considering that the government can arbitrarily set both crop prices and other input prices, it is not even clear that such an irrigation fee would achieve any fiscal benefits to the government. If the fee were eliminated, the government could maintain its revenues simply by changing one or more of the other prices used in its calculations.

11.3.4 *Avoiding excessive collection costs*

Enforcing the collection of irrigation fees is important. Not only does it generate current revenue, but it also maintains the integrity of the system of fees which is critical to the long-term ability to generate revenues from the fees. But enforcing the collection of fees can also be expensive. It is important to find enforcement methods that can be effective without incurring too much additional cost. In other words, in enforcing irrigation fees, as in other types of economic activity, we must seek cost-effective methods.

The cost effectiveness of enforcement methods must be continuously re-examined. A method that may be quite costly to implement may be cost effective at a point in time where the collection rates would otherwise be very low. But the same method may not be cost effective at a later point in time when familiarity with the concept of paying for irrigation services is greater, and the level of payment that would exist in the absence of this method would be higher. An example of how cost effectiveness may change over time is given in Box. 11.9.

11.4 *Summary*

A system of user fees can be sustained only if reasonably high rates of collection are obtained. Not only must conditions be created that encourage individual water users to pay their fees, but the irrigation

---

BOX 11.9

Collecting irrigation fees in kind in South Korea

In the early 1950s, irrigation associations in Korea were having difficulty collecting payments for irrigation services from farmers in a timely fashion. To alleviate this situation, the government formulated new regulations in 1952 that allowed the irrigation associations to collect the payments either in cash or in kind. The fees continued to be assessed in cash, but payment in kind could be made by converting the assessed value to cash at the market price of paddy prevailing at the time of the assessment of the fees. Approximately half of the irrigation associations in South Korea chose to take advantage of this new option and collect the irrigation fee in kind. This procedure resulted in considerably improved rates of collection.

But two problems were encountered with the payment in kind method. First, the National Agricultural Cooperative Federation (NACF), which eventually acquired the grain delivered in payment of the irrigation fees, found itself with varying amounts of several different grades and varieties of paddy. Second, variations in the moisture content of the paddy received from farmers introduced problems in the handling and post-harvest processing. As a result of these problems, losses were incurred by the county branches of the NACF.

In spite of these difficulties, payments in kind continued into the 1970s, with about 80% of the total value of fee collections in 1975 derived from collections in kind. But by this time collection rates were virtually 100%, and collection in kind no longer served a purpose of improving the rate of collection. As the costs of collecting in kind began to outweigh the benefits, more emphasis was placed on collection in cash. By 1980, only 20% of the collections were in kind. By 1984 collection in kind had been eliminated, with farmers required to pay their irrigation fees in cash.

---

agency needs incentives to take positive actions to obtain high rates of fee collections.

Factors that can encourage users to pay fees include:

(i) giving the water users a perception that they own the irrigation facilities;
(ii) making the personnel of the irrigation agency accountable to the water users for the performance of the irrigation system;
(iii) earmarking fees collected in a particular project for use in the operation and maintenance of that project;
(iv) paying water users' organisations a commission to collect the fees from its members;
(v) providing positive financial incentives, such as discounts for prompt payment;
(vi) providing positive social incentives for payments;
(vii) making direct contact with the water users, rather than waiting for the water users to come to pay their fees;
(viii) giving responsibility for collection to intermediaries who are in more direct contact with the water users; and
(ix) allowing the users to pay their fees in kind rather than in cash.

In addition, a system of enforceable penalties for non-payment of fees is needed. Termination of water deliveries is potentially the most effective penalty; however, for a variety of technical and political reasons it is not often used. Financial penalties can be imposed on those who pay late, and social sanctions may often be effective. Legal penalties sometimes suffer from being so cumbersome that even the threat of their use carries little credibility; however, in some cases simplified and very effective legal procedures have been developed specifically to deal with irrigation matters.

An irrigation agency may be able to facilitate the collection of irrigation fees by linking them with other charges (such as land taxes or charges for fertilisers or other inputs provided by the government). The danger of this approach is that by making the collection of the fees nearly automatic, the linkages that a system of user fees can create between the water users and the irrigation agency may be severely weakened. As a result, the system of fees may not be able to create the various positive effects on irrigation performance that we have discussed in previous chapters.

Although collection of the fees is important, an irrigation agency must not lose sight of the fact that collection is also a costly process. It is important that the cost effectiveness of the collection methods be evaluated periodically in light of changing conditions.

12

The political economy of irrigation financing

This chapter will outline some of the political questions relating to irrigation financing. There is a variety of political interests and pressures affecting the practical establishment and implementation of financial reform of irrigation institutions. Inevitably the issues that arise will include the political viability of any scheme for irrigation fees; the genuine political problem of charging low income farmers anything at all, even if product prices are increased; the opposition to reform of officials who benefit from the opportunities for corruption created by the economic rents that occur in subsidised irrigation; and coping with the genuine social and cultural impediments to change when traditional values are in conflict with modern methods of financing.

Prospects for reducing political difficulties with a user fee policy clearly have to be explored. In particular, we have discussed the idea that decentralised and financially autonomous irrigation organisations could reduce the political visibility of irrigation fees. The importance of ideas such as indexing fees to maintain their real value over time, thereby reducing the frequency with which changes in the rates need to be made, also needs to be considered. The possibility of using the changes created by improvements such as rehabilitation as an opportunity for reforming the system of user fees will be reviewed. But first the general area of the politics of water must be explored.

12.1 *Water charges and culture*

Questions such as whether, how, when and at what level user fees are to be charged are matters that must take full account of the surrounding culture, rural social and institutional structures, and behavioural 'norms'. As noted in the quotation in Chapter 1 from Starr and

Stoll, strong cultural attitudes about water and water rights exist throughout the world. Reformers must have a proper understanding of these issues and a sensitivity to the way in which institutions are evolving. On the other hand, it must be recognised that politicians often exploit an idealistic view of cultural 'norms' about charges for water as a self-serving method of justifying a popular political stance of providing irrigation water to the users free of charge. In one country we visited, for example, we were told that 'all water must be free in our culture'. Yet on the ground in the various countries we visited, we consistently found that farmers were faced with a battery of official and unofficial levies of money, labour and indirect taxes that showed irrigation to be far from free. Even in some cultures where religious norms specify that water is a free gift of God, fees for the provision of irrigation services have been found acceptable.

Nevertheless, reformers pushing for increased financial autonomy must be aware of the influence of strongly held traditional attitudes and values about access to and charges for water that make adjustments in financing arrangements difficult to achieve. We firmly believe that reform of underfunded irrigation institutions providing a substandard, unreliable, often deteriorating service is in the direct interest of all parties – farmers, irrigation staff, governments and consumers. In our view, it is this opportunity to provide benefits to all (an opportunity that is rare in development practice) that is the potential foundation upon which political acceptance of financial reforms may be built.

12.2 *Water politics*

12.2.1 *Sectoral issues*

Political resistance to establishing or increasing irrigation fees can be anticipated if irrigation appears to be treated differently from other subsectors within an economy. This will be the case whether or not general macroeconomic disequilibria make widespread reform desirable. If financial reform in irrigation is to succeed there should be no suggestion that irrigation is to be a 'vanguard' sector or the 'cutting edge' of structural adjustment.

If, for example, urban folk do not pay, in full, their drinking water fees, their telephone bills or electricity charges, or if they face bills that are subsidised, then why, it will be argued, should increased irrigation fees be imposed on poor farmers. Similarly if farmers are impoverished by overvalued exchange rates that heavily subsidise the mainly urban importers and tax agricultural exports then further impoverishment through higher water charges is clearly not to be recommended.

We thus see the arguments for water institution reform as first and foremost part of the general process of adjustment and regeneration in a distorted economy.

12.2.2 *Politics and ethics*

In the end all practical or applied economics is an exercise in political economy which is concerned with scarcity, choice, trade-offs, efficiency and the like within a framework of power, pressures and notions of fairness. As John Maynard Keynes said, 'economics, more properly called political economy, is on the side of ethics'.

We contend that analysis of the ethics of irrigation reform and options facing irrigation farmers will show:

(i) that it is generally rational for farmers to contribute collectively more resources than at present to maintain the systems from which they benefit;
(ii) that nearly all participants (farmers, system employees and consumers) could mutually benefit from reform; and
(iii) that such reform would be equitable as subsidised irrigation farmers may be poor but they are seldom the ultra-poor.

Furthermore, our field investigations show that farmers are often willing for government enforcement of existing or even higher fees because they recognise that both their collective and individual interests in irrigation are served by strong management; however, they often have great doubts about the willingness and competence of the State to fulfil promises and deliver honest, proficient and reliable resources. They are properly unwilling to pay for substandard service.

12.2.3 *Conflicts between collective and private interests*

These points take us into a contentious area of public choice theory. Individual irrigation farmers generally recognise that the collective interest is served by strong management that enforces rules. Furthermore, they understand that if all follow these rules their aggregate individual benefits are maximised. However, large potential rewards exist for any one farmer who can escape the rules (e.g. by not paying the irrigation or drainage fees; by overexploiting a limited groundwater source; or by stealing extra canal water at peak, high-value periods) if all others continue to follow them. We thus get the paradox of the irrigation farmers longing for rules and strong management to enforce them, yet following patterns of individual behaviour that are incompatible with these goals. Analogous situations include fishermen who overfish a lake, herdsmen who overgraze a common pasture, and some would argue, families who have more children than the individual and society can educate, house and provide with health care. In our field experience most irrigation farmers long for strong management but by their actions they undermine the possibility of achieving it.

12.2.4 *Sources of pressures for reform and the role of donor agencies*

The dynamics of water politics are strongly influenced by the sources of pressures for reform. The strongest voice for the imposition of user charges for irrigation frequently comes from bilateral and multilateral donor agencies. Governments may find that nominal acceptance of the positions advocated by the donor agencies is a price that must be paid for the aid package. But without a strong internal political commitment to these policies, the reform efforts easily collapse upon encountering even modest domestic opposition. The relatively unsuccessful attempt in Sri Lanka in the mid-1980s to collect an 'operation and maintenance fee' in irrigation projects is a case in point. The impetus for establishing the fee came primarily from the World Bank. Responding to this pressure, the government attempted to implement the fee quickly, before thoroughly working out a number of critical details. This in turn led to difficulties in implementation that provided opportunities for political opposition to the policy to coalesce.

Pressures for the implementation of irrigation fees have consistently come from the World Bank, which generally incorporates cost recovery covenants into its loan arrangements for irrigation projects. The failure of this pressure to result in significant changes in policy is indicated by a study undertaken by the Operations Evaluation Division of the World Bank in the mid-1980s. The study found widespread lack of compliance with the cost recovery covenants of the Bank's loan agreements.

This failure of the pressure from external donors to result in changes in policy can in part be attributed to the fact that the project process of donor agencies such as the World Bank generally does not involve discussions with politicians. During such key stages as project appraisal, the staff of the donor agency have discussions with technical people in ministries of agriculture, irrigation and finance, but seldom have contact with the politicians who must ultimately bear the political risks of policy changes. Because of this gap in the appraisal process, the donor agency is unlikely to fully comprehend the political realities surrounding proposed policy changes, and host country politicians are unlikely to understand fully the importance of suggested reforms.

Another common problem resulting from external donor pressures for reform is inconsistency between policy objectives and policy actions. A donor agency, which at its policy-making level is concerned with the sustained funding of irrigation operation and maintenance costs, may insist on the implementation of cost recovery through user fees. The host country may reluctantly agree to a nominal cost recovery effort; however, if, as is frequently the case, the country has no tradition of financial autonomy for irrigation, there is likely to be no effect on the funding of irrigation operation and maintenance. The operational office of the donor agency is likely to accept this compromise position, as it allows the project to proceed within the host country, while meeting, at least nominally, the requirements imposed by the policy arm of the donor. The result, however, is a no-win situation: the government is unhappy because it has been forced to impose what is generally seen as an unwanted additional taxation burden on the irrigation farmers; and ultimately the donor agency will be unhappy because the long-term problems of funding irrigation operation and maintenance have not been seriously addressed.

True policy reform in irrigation financing is much more likely to take place when the pressures for reform are internal. In the Philippines, for example, severe budgetary pressures on the central government led to a decision to terminate, over a five-year period, the government funds provided to the National Irrigation Administration for irrigation operation and maintenance. Accompanying this decision was a change in the NIA's charter that allowed it to retain funds collected from irrigation fees (see Box 9.6 in Chapter 9). For the NIA, the financial autonomy which it obtained by this change provided a powerful incentive to try to improve the effectiveness of the system of irrigation fees. Furthermore, because these policy changes were initiated on the basis of domestic rather than external political pressures (although undoubtedly at the macroeconomic level external pressures played a significant part), the political commitment to their successful implementation has been far greater than in the Sri Lankan case cited above.

Developments over the last 30 years have extended the scope of government activity, direction and control within most countries. We are presently in an era in which many of these aspirations are seen as having been excessively ambitious, resulting in the over-extension of government activities beyond the nation's resources and management capacity.

Many governments, having lost confidence in their ability to enforce policy, rules, laws and taxes, are seeking new ways of thinking to lessen the financial and administrative burdens that these policies have created. These domestic political forces provide opportunities to consider reforms that include looking to self-regulating market mechanisms such as cooperative ownership and management of systems, full privatisation, and contracting out of management services.

12.2.5 *Reducing the political visibility of user charges*

As noted in Chapter 10, regular adjustments in the nominal monetary value of service fees are essential if their real value is not to be eroded. But making frequent changes in the fees is often difficult because of the political visibility of the decisions to make these changes. This is particularly true when a uniform fee structure exists for an entire nation, so that the decision to change the fee is a national decision. Changes in prices by administrative decision can bring citizens into the streets in protests or even riots.

We support the idea of fees being specified in commodity terms, such as say 100 kg paddy or wheat of a particular grade per hectare per season. There is no need to actually have the grain delivered as payment for the water fee; the establishment of the cash equivalence is fully adequate as a means of building an index against inflation into the fee structure. This index minimises the risks associated with inflation and uncertain, intermittent increases in monetary fee levels, and yet it is easily understood by everyone. Such an indexing policy does carry the lesser risk that a glut of the indexed commodity or long-term decline in its relative price will reduce the value of revenue.

An institutional setting of decentralised financial autonomy also helps reduce the political visibility of irrigation fees. When any given irrigation agency must increase the level of fees, the effects are felt only by the farmers served by that agency, rather than by farmers throughout the nation. Furthermore, there is likely to be better communication between the farmers affected by the decision and the agency making the decision than would be the case where a centralised agency makes a decision affecting irrigation fees throughout the nation.

12.3 *Lessons from land reform*

We might learn something of relevance for water policy reform by looking at the recent history of land reform. In the 1960s land reform was a popular area for political debate and for public policy; however, it faded from prominence in the 1970s. Although a variety of reasons for this change can be cited, one significant reason was the resistance of those who were threatened with the loss of assets and related power. Land reform might provide benefits to many, but only at the cost of large losses to some.

What are the implications of this land reform experience for water reform? First we should not underestimate the political problems of achieving changes and the strength of resistance to plans for reform on the part of holders of assets who are already achieving significant economic rents. But even more importantly, ways should be sought to ensure that the reforms do not create net losses to the irrigation farmers. The key is to devise water fee reforms that provide financial resources to improve irrigation operation enough so that most or all of the farmers will find themselves better off under the financial reform policies.

Farmers, however, have a natural distrust, often reinforced by experience, of government promises that implementation of a system of user fees will result in improved services from the public sector, leading to higher incomes and increases in land values. Their experience suggests that governments do not, perhaps cannot, deliver.

The logic of these considerations (a logic that is as yet relatively untested by experience) suggests that the proper time to bring in a water fee reform is as part of a modernisation or rehabilitation programme. This provides an opportunity to establish a contractual agreement between farmers and government that publicly describes the obligations of both parties. The timing of water fee reforms and the sequencing of changes is crucial to success. We believe that the most appropriate time to discuss proposals to increase fees is during identification of modernisation or rehabilitation projects. Only when a demonstrable improvement in service is to be made can agreed fee increases be introduced with confidence.

One implication of this approach is that it eliminates the possibility of a uniform national irrigation fee. A nation could, however, establish a uniform base fee to which a modernisation or rehabilitation premium would be added on a project-by-project basis as and when the investments were made. It is likely that over the next 25 years most of the world's irrigation will receive some modernisation investment that will increase either the availability of water or the reliability of its delivery. Especially in nations lacking a tradition of significant user fees for water, we feel that the ideal time to bring in new fees, locally retained and locally administered, is when this modernisation investment occurs.

12.4 *Briefing politicians*

Policy reform in relation to irrigation financing requires political support. Convincing politicians of the need for reform is thus critical if the reform process is to proceed. What are the steps in briefing a politician on the need for a water reform? How can water specialists convince the sceptical and sometimes cynical politicians that water fee reform is worth the fight?

The first step is to establish the financial value to a farmer of a reliable irrigation service over and above the next best alternative. Depending on the existing situation, this alternative may be agriculture either under rain-fed conditions or with poorly managed and unreliable irrigation, both of which are more risky and less profitable than farming with reliable irrigation. Clues to the likely level of the increments in income due to the provision of irrigation can be obtained from farm economic surveys of irrigated and rain-fed farming. Expected increments in income due to improvements to make an irrigation system more reliable can be established by looking at private irrigation or tubewell irrigation from groundwater or even by studying the farms at the heads of canals and watercourses where water supplies are usually in good order. Where tubewells are privately owned and the full costs are borne by farmers, they often pay for capital and recurrent costs of water at substantially higher levels of fees (often 10 or 20 times higher) than the fees charged for public sector surface irrigation water. This information alone is an important indicator of the economic value that farmers obtain from reliable irrigation.

The second step in briefing politicians requires an assessment of the economic value, to the nation, of irrigated farming. This step requires the shadow pricing of inputs and outputs. The purpose of this analysis is to demonstrate to the politician the true economic or social value of irrigation and the impact of the proposed charges. Consider the hypothetical example in Table 12.1. In financial terms, a typical farmer can expect a gross income of Rp 1000 with a margin after all costs of Rp 650; however, the economic analysis shows that the net value to society of this production is only Rp 500. In the economic analysis the gross income is increased to adjust for an overvalued exchange rate that 'taxes' crop exports. In spite of this, the adjustments of the subsidised input costs to economic prices and the inclusion of the true costs to society of the irrigation input reveal that at current levels of productivity the net economic value is less than the market value earned by the farmer.

Many economists would be urging the politician to move all distorted market prices toward economic prices or social opportunity costs. This

Table 12.1. *Hypothetical income and costs per hectare (Rp)*

Item	Market prices	Economic (shadow) prices
Gross income	1000	1300
Variable costs (excluding irrigation)	300	500
Gross margins (excluding irrigation)	700	800
Irrigation costs	50	300
Gross margins or net revenue	650	500

would have the effect of raising prices received by farmers for their output, and eliminating subsidies they receive on the inputs they use. In the example in Table 12.1 this would result in a sixfold increase in irrigation costs and a lower net return to farmers. It would require a very well-established and authoritarian politician to bring in such a bold reform.

What our analysis of the example of Table 12.1 lacks at the moment is a recognition that if farmers paid a higher proportion of the true costs of irrigation, and if these funds were devoted to improving the present unreliable irrigation service, then yields (and crop area) could increase substantially. Except in the most effectively operated irrigation systems, we are convinced that with enhanced accountability linkages and increased funds targeted to improve irrigation service, gross returns would be increased substantially (by as much as double in some cases) without any change in any other inputs. In our example, a doubling of the gross returns as a result of improved irrigation would cause the economic gross margin (that is, with all costs paid in full) to increase from Rp 500 to something over Rp 1500 depending upon the amount of the increased expenditures on irrigation.

Our example was chosen to illustrate the prime importance of obtaining increased production stemming from the increase in fees, and the allocation of the resulting funds to improve the volume and reliability of water supply. Its validity is crucially dependent on two assumptions: that a combination of underfunding and the associated weak accountability linkages between the irrigation system operators and the farmers substantially reduces the quality and quantity of service, and that user fees would be collected and devoted to service improvements. We believe the former assumption is a realistic assessment of the situation in many irrigation projects. The latter assumption involves a matter that is subject to policy decisions.

The third step is to convince politicians of the critical importance of a locally levied and locally retained water charge. This is perhaps one of the most difficult steps because in many cases it will imply the need for significant institutional reform in the direction of financial autonomy. Yet, as illustrated in the example discussed in the previous paragraphs, it is precisely the benefits of financial autonomy, in the form of increased accountability and increased availability of funds for O&M, that are the key elements of the policy reform that allows an increase in user charges to be linked to an increase in the economic and financial returns to the water users. And this, in turn, is the key political factor that allows the policy reform to be seen as creating positive benefits for farmers, rather than as the imposition of another burdensome tax.

The fourth step, closely related to the third, would be to find some institutional mechanism whereby the irrigation water customers can engage and manage their suppliers in the irrigation department. This could range from informal groups where water user association representatives debate with those that supply water, to a system where all employment of irrigation department personnel is in the hands of farmers. Greater accountability to customers is crucial. It has to be recognised that unless farmers first believe that extra fees will improve their irrigation, and then see that there is an improvement that is sustained, they either will not pay their fees, or will soon cease payments.

It is ironic that on those projects where corrupt officials extract bribes from farmers for water there already exists what is in effect a locally retained fee for irrigation, with local accountability to deliver the promised and paid-for service. What we are advocating in this book is in a sense the legitimising and institutionalising of a commonplace but corrupt system, and thereby making it possible to use the fees for socially productive ends. We do not underestimate the difficulties of undertaking such institutional changes; however, we believe that reform of corrupt systems is possible, particularly where strong economic incentives for change exist.

The fifth step is to establish practical contingency plans for waiving the fees and for financing irrigation in the event of a natural disaster or an unanticipated fall in farmer income. The farm production or income 'norms' that would trigger these exemptions should be set in advance, so that these plans are in fact used only on those rare occasions when they are truly needed. A contingency plan of this type has been established by the National Irrigation Administration in the Philippines. It provides a schedule of fee reductions related to the degree of shortfall in crop yields caused by natural disasters such as typhoons. In cases of very large reductions in yields, the entire fee is waived.

Finally, politicians need to recognise the existence of several positive but broader implications of enhancing the efficiency of irrigation service. The availability of profitable complementary agricultural technologies associated with new crop varieties has increased the potential social returns to irrigation, thereby raising the opportunity cost of inferior irrigation services. Likewise, these opportunity costs have been increased by the rising needs for agricultural products stemming from growth in population and income, with their resulting pressures on the more fragile rural environments. Efficient irrigation has the potential to create significant environmental benefits by establishing a substantial agricultural system that can draw people and animals away from the more fragile rural environments. As politicians come to understand the linkages between policy reform in irrigation financing and the realisation of these broader environmental and social gains, the prospects of generating the political support necessary for the implementation of the policy reforms are enhanced.

Politicians will need to be convinced by their economic advisers, and we hope by the arguments in this book, that direct cost recovery from enhanced user fees that are retained on the scheme to finance operation and maintenance costs is economically efficient, equitable, fiscally efficient (remembering that subsidies must always be paid for by someone somewhere within the economy), and, potentially at least, organisationally feasible.

12.5 *Mechanisms and goals for irrigation financing and cost recovery*

For existing irrigation projects, we feel that a reasonable practical goal is to obtain, through the cost recovery mechanism of user fees, resources to be used to finance irrigation operation and maintenance at the economically appropriate level, which is usually higher than the present level of expenditures on these activities. This is a very low-level ambition in gravity schemes where a high proportion of costs are capital costs. This goal is advocated on pragmatic grounds, in that it is readily understood by all, and in most countries it will result in substantial increases in revenue.

We prefer user fees for O&M costs to other burdens that might be imposed on farmers, such as product levies, export taxes, increments in land taxes, or delivery quotas of key commodities. Once again we are led to this conclusion on pragmatic grounds. In particular, these mechanisms fail to create the accountability linkages that a system of user fees can foster. Furthermore, some of these mechanisms tend to discourage production, and they seldom result in revenues being actually used to finance the costs of irrigation.

For proposed new investments to enhance irrigation, such as electrification of pump schemes, conjunctive-use projects linking surface and groundwater, canal lining, pipe distribution, or application with sprinkler or trickle technology, we suggest initial consideration of the twin goals of attempting to recover the full capital cost through a betterment levy to be paid by the landowner, and of financing the full recurrent costs through a user fee. These plans for cost recovery and financing need to be made an integral component of the investment decision process, so that the landowners and water users would be encouraged to weigh carefully the costs of the improvements against the likely benefits. Of course, full cost recovery may be neither possible nor desirable if significant distortions exist in the pricing of important inputs or outputs related to the investment, if significant externalities (such as downstream drainage benefits) exist, or if important elements of collective goods (such as enhanced food security) exist. Furthermore, as noted in Chapter 10 (see Box 10.2), if the variability of the benefits among the irrigated farmers is high, the need to set the fee at a level low enough to ensure that for most farmers it is less than the benefits they receive, may significantly reduce the level of cost recovery that is possible.

The goals suggested above can serve only as general guidelines. Each case must be established on its merits and there will always be circumstances where little if any cost can be recovered. Drainage is a good example of a special case. There are some irrigation schemes where expensive drainage appears essential to maintain current production. But why should today's cultivators pay for neglect of drainage and poor irrigation practice in the past which caused the current drainage problem, and why should farmers at the tail of canal systems be charged to remove salt, part of which has been dumped in their irrigation water by upstream farmers? Such problems need critical examination and the incidence of costs and benefits needs to be carefully established.

In the final analysis the neglect of irrigation potential has a real or opportunity cost, and yet any investment has to be financed in some fashion. The strong golden rule that will have to be tested is that the water users and landowners who directly benefit from irrigation should bear the cost of this production input. At present in most circumstances payment is being forced on someone else, with the concomitant result that the amount being paid is not enough to provide a high quality of irrigation service.

13

Conclusions and recommendations

As irrigation development in the Third World has proceeded over the past several decades, two problems have emerged. The first is an increasingly severe shortage of funds to operate and maintain the irrigation facilities and to rehabilitate and modernise old systems. The need for such funds has grown in conjunction with the increase in irrigated area resulting from the continued investment in irrigation. As the fiscal and debt problems of the 1980s have thrust themselves on many Third World nations, governments have encountered particularly severe difficulties in providing the needed funds.

The second pervasive problem of irrigation has been poor overall performance relative to expectations. These performance difficulties are generally attributed to deficiencies in irrigation management. Many of the deficiencies involve unsatisfactory procedures for operation and maintenance; however, some of these problems can in turn be traced to deficiencies in the investment process of planning, design and construction.

These two problems are related in one obvious way: inadequate funding for recurrent costs may make it impossible for an irrigation agency to operate and maintain the facilities in a satisfactory fashion. But the two problems are also related in a more subtle fashion. A government's policies for irrigation financing can affect the relationships among the various actors of an irrigation system and have a profound effect on their behaviour. Methods of financing recurrent costs of irrigation can thereby affect the quality of irrigation management and this in turn affects farmer investment and field management.

The focus of this book has been on examining irrigation financing policies with a view to considering both how well they provide the funds needed to operate and maintain the facilities and how well they encourage good management so that the funds obtained are used in ways that provide an effective and efficient irrigation service. We have evaluated policies on the basis of the criteria of resource-mobilisation efficiency, resource-use efficiency (including effectiveness of operation and maintenance, effectiveness of investment decisions, and water-use efficiency) and equity.

Our preferred approach to financing the costs of irrigation can be stated simply: implementation of user fees by a financially autonomous irrigation agency. This preference stems from our analysis of irrigation financing policies in a number of nations around the world in the light of the above efficiency and equity criteria.

From an efficiency perspective, a policy of user fees implemented by a financially autonomous irrigation agency creates the potential for improvements both in the operation and maintenance of existing irrigation facilities, and in the process by which investment decisions are made.

The potential for improvements in operation and maintenance stems in part from the greater control that a financially autonomous irrigation agency can have over its budget. Rather than relying on a potentially arbitrary and capricious budgetary process of the central government, a financially autonomous irrigation agency has an ability to establish the user fees at levels that should provide it with the needed operating funds. But the necessary revenues will be forthcoming only if the agency is able to collect the fees from the water users. This need of a financially autonomous irrigation agency to secure the cooperation of the users with respect to fee payments facilitates the creation of accountability linkages between those who operate the irrigation facilities and the farmers. Because these linkages increase the likelihood of the needs and perspectives of the water users being taken into consideration by the irrigation agency, the potential for improved irrigation performance is established. These accountability linkages may also give water users a voice in determining how the irrigation agency's funds are to be used, thereby involving them in the process by which the total 'need' for funds is determined.

The potential for improvement in the investment decision process resulting from the implementation of user fees by a financially autonomous irrigation agency stems from the fact that the agency must set the user fees high enough to cover all or part of the costs of irrigation resulting from an investment decision. This gives the agency a direct vested interest in the economic and financial viability of proposed investments.

Contrary to many others, we place relatively little importance on the possibility that user fees will encourage individual farmers to be more efficient in their use of water. This is because fees are likely to affect a farmer's decisions about water use only if they are structured as water prices – where the total amount that must be paid for water depends on the farmer's water-use decisions. But for a variety of reasons, including the physical difficulty and cost of measuring water deliveries to large numbers of small farmers, most systems of user fees in Third World countries are presently structured so that a water user's payment is fixed (usually on some basis related to the area irrigated) irrespective of the amounts of water actually used. Under these conditions the fee becomes a fixed production cost to the farmers, and therefore does not provide an incentive for them to economise on the use of water.

The criterion of resource-mobilisation efficiency emphasises the need to minimise the social costs of acquiring the funds to pay for irrigation. It is difficult to make generalisations about user fees with respect to their resource-mobilisation efficiency. Compared with financing methods that involve general government revenues collected through an existing tax system, user fees are likely to result in higher administrative costs but lower economic distortion costs. Administrative costs of a system of user fees can be minimised by keeping the fee structure simple. This is one reason why user fees are often based on the area irrigated, rather than on a more complex system of volumetric water prices. A significant portion of the administrative costs of user fees does not vary with the amount of funds actually collected, making the rate of collection of the amounts assessed a key factor affecting the efficiency of resource-mobilisation. If rates of collection are extremely low, administrative costs can actually exceed the amounts collected.

User fees also have certain advantages from an equity perspective. First, they can reduce the need for government subsidies. Because these subsidies tend to benefit large farmers more than they do the small and poorer ones, their reduction and replacement by a fee that tends to vary in proportion to farm size would be viewed by many as a change that makes irrigation financing more equitable. Second, even though water users are often very poor, they are likely to be better off than those who farm under rain-fed conditions or the increasing number of rural landless. By requiring the water users to pay for at least part of the costs of irrigation, the government should be in a better position to undertake development projects that would benefit the even poorer rain-fed farmers.

Although we generally favour a financial policy of user fees under financial autonomy, one clarification and several qualifications about this preference must be noted. The clarification is that our preference for user fees is contingent on the existence of financial autonomy for the irrigation agency. Only in the context of financial autonomy can user fees be expected to result in the efficiency benefits discussed above. In the absence of financial autonomy, user fees become just another tax that is not necessarily superior to other taxes that might be levied to finance irrigation.

The first qualification regarding our policy preference is that in some situations we would find a policy based on financial autonomy and user fees to be undesirable. A government's policy toward irrigation financing is only one of many government policies affecting the agricultural sector. It is therefore inappropriate to consider irrigation financing policy in isolation from the broader macroeconomic environment in which it will be implemented. A nation's tax structure, its exchange rate, and the relative prices of agricultural and non-agricultural goods are all determined by a complex set of policies. If, as often happens, this combination of policies places a heavy implicit tax burden on the agricultural sector, the introduction of a system of user fees for irrigation might be inappropriate. It is even possible that in such a situation the government would provide subsidised irrigation services with the specific objective of offsetting some of this tax burden.

The second qualification is that a system of user fees under financial autonomy does not necessarily imply that the users must pay for the full cost of irrigation. In some cases asking users to pay for the full costs would be unreasonable because the costs of irrigation are too high. High costs sometimes reflect inefficiency and corruption; however, sometimes they simply reflect the fact that political or social factors overrode narrow economic considerations in the initial decision to build an 'uneconomic' irrigation project. It would also be inappropriate to ask the users to pay for the full costs of irrigation in situations where irrigation involves significant externalities, or where it results in outputs with the characteristics of public goods. Another reason why it may be unreasonable to charge farmers for the full costs of irrigation is the existence of a heavy implicit tax burden on farmers resulting from the general taxation and price policies discussed in the previous paragraph.

Even if user fees are less than the full costs of irrigation, they can still provide many of the benefits discussed above. In particular, most of the efficiency benefits of user fees under financial autonomy can be obtained even if the fees cover little or none of the investment costs. Once an irrigation project is constructed, a government's decision to subsidise part or all of its construction costs is a decision regarding the financing of a cost that, because it occurred in the past, is sunk. The benefits of user fees under financial autonomy arise, however, not because they recover sunk costs, but rather because of the linkages they create between the water users and the irrigation agency with respect to current and future costs. It is therefore more important that user fees cover recurrent costs than initial construction costs.

Our third qualification involves situations where tenancy is widespread. In these situations, the direct benefits of irrigation are likely to be shared between the actual water users (many of whom are tenant farmers), and the landowners (many of whom do not actually engage in farming). The water users should receive some increase in their annual incomes as a result of irrigation, while the landowners can be expected to reap an increase in the value of their land, which may be reflected in part by an increase in the annual rental payments they receive from their tenants. In these cases where a significant sharing of the direct benefits of irrigation between water users and landowners exists, it may be desirable for a nation's irrigation financing policies to involve a combination of user fees and a land tax or a betterment tax levied on the owners of the irrigated land.

The fourth qualification regarding our irrigation financing policy preference is that no matter how desirable a system of user fees under financial autonomy might be in general, it may be extremely difficult for a government to make an irrigation agency financially autonomous in a specific situation where a tradition of financial autonomy is lacking. Significant institutional changes in areas such as administrative rules and procedures may be required to introduce financial autonomy. This does not necessarily mean, however, that nothing can be done. Where such changes appear to be daunting, it may be possible to move gradually towards financial autonomy. Earmarking of irrigation fees for the budget of the irrigation agency can be a major step in the direction of financial autonomy.

Our final qualification about our preference for user fees and financial autonomy is simply to note that a careful assessment should be made of the political feasibility of implementing such an irrigation financing policy. Strong political opposition to user fees can be expected in some nations. Unless there is reason to believe that a government has the political strength to withstand such opposition, there is no point in attempting to introduce such a policy. However, in most circumstances we believe the long-term economic gains from reform will outweigh the short-term political costs. We are, of course, writing as professional economists and not as practical politicians!

Not everyone will agree with our general presumption in favour of an irrigation financing policy of user fees under financial autonomy, even after taking account of the above qualifications. Opposition to user fees tends to be stated in terms of equity concerns. Two of the most common equity arguments given against user fees deserve further examination.

The first argument focuses on the fact that water users are not the only beneficiaries of irrigation. Firms that supply agricultural inputs and those that market the irrigated crops benefit from an increased volume of business. Consumers often benefit through lower prices. User fees, which force water users but not any of the other beneficiaries to pay for the costs of irrigation, are therefore seen as inequitable.

We already have examined the strengths and weaknesses of the argument in some detail in Chapter 8. The primary point that we wish to make here is simply that this is really not an argument against user fees. Rather, it is an argument against trying to cover the full cost of irrigation through user fees. It is thus possible to accept this argument while still agreeing with our position favouring user fees to cover the costs of O&M.

A second equity argument commonly raised in opposition to proposals for user fees is that because many of the farmers who benefit from irrigation projects are very poor, equity is served by having irrigation subsidised from general government revenues, which are derived from taxes on wealthier people. Unlike the first argument discussed above, this argument can be used to suggest that no user fee should be imposed for irrigation.

This equity argument is a reflection of a broad concern about income distribution and social justice. But it is our observation that irrigation financing policies are not very good or powerful tools for dealing with these concerns. Failing to charge users for irrigation water will have an effect on income distribution that is small relative to the effects of the general economic policies that determine input and output prices and exchange rates. Furthermore, free irrigation water has the perverse equity effect of benefiting larger farmers more than small ones. In our view, irrigation financing policies will be more valuable if they are designed to meet effectively their primary objective of mobilising resources and establishing accountability linkages that help ensure a well functioning irrigation system. A system of user fees could incorporate some safeguards for severe hardship cases (such as exempting from the fees a specified minimum acreage for each water user); however, in general, legitimate social concerns regarding income distribution can be addressed much more effectively by other types of policies.

In reality, however, this second 'equity' argument is often more of a political argument against the imposition of user fees than it is a clearly reasoned equity argument. User fees, particularly in situations where irrigation responsibilities are centralised, have a high degree of political visibility. It is relatively easy to make a political case against user fees on the grounds that the water users are poor. The alternatives to user fees, such as increases in general taxes, reductions in other government expenditures, inflationary financing, or deterioration in the irrigation infrastructure and in the quality of irrigation services are likely to carry much less political visibility with respect to irrigation, either because they are not specifically linked to decisions about irrigation financing or because their effect is not immediately apparent. As a result, the alternatives to user fees are politically much easier to implement.

Having made (with qualifications) the case for user fees implemented by financially autonomous irrigation agencies, we turn finally to address questions surrounding the key implementation matter around which a system of user fees is likely to stand or fall, namely, the ability to collect the fees from the water users.

Financial autonomy for irrigation agencies cannot succeed unless these agencies are able to collect a reasonably high percentage of the fees assessed. Governments may be reluctant to implement financial autonomy for irrigation because of fears that an inability to achieve satisfactory rates of collection could lead to a deterioration in the irrigation agency's financial situation, with serious negative implications for the nation's irrigation infrastructure.

Although there is no simple solution to the problem of fee collection, we believe that the prospects for success can be enhanced by giving careful attention to three considerations. The first and most fundamental consideration is equity: establishing a basis for assessing the fees that is deemed equitable by the users. The second consideration is the establishment of positive incentives that provide encouragement to water users to pay their fees. Finally, effective penalties for enforcement of the system of fees are needed.

We have already emphasised the subjectivity of equity as a criterion for the evaluation of financial policies. But in establishing a system of user fees, the chief equity concern should be the perspective of the water users themselves. If there is a general perception among the water users that the fees are assessed in an inequitable manner, it will be difficult or impossible to obtain satisfactory rates of collection unless enforcement methods are extremely strong. This is one reason why water users need to be involved in the design of a system of irrigation fees.

The equity concerns of water users are likely to pertain to their perceptions of important differences among themsleves in terms either of the typical benefits that they receive from irrigation, or of the costs that must be incurred to serve them. Some of these differences in benefits and costs may be relatively permanent. This would be the case when they stem from differences in the physical characteristics of farms (such as size, topography, or soil texture), or from the history of the development of the irrigation facilities (such as where an investment to increase the water supply permitted an irrigation system to be expanded to serve a larger area). But some differences may be transitory, reflecting variability in conditions from year to year. Provisions for the reduction, postponement or forgiveness of the fee in cases of severe crop losses due to pests or natural disasters such as typhoons could emerge from these concerns.

As the above examples suggest, the need to establish a system of fees that is equitable in the eyes of the water users may result in a fee structure that is more complex and thus more costly to administer than would otherwise be the case. Careful attention needs to be given to the trade-offs between administrative costs and equity. But the crucial importance of the users' perceptions of equity to the viability of the system of fees suggests that some increases in administrative costs to accommodate equity concerns may be desirable.

An equitable basis for fee assessment provides the foundation for successful implementation of a system of user fees. But success requires more than a foundation. Ways need to be found to provide the users with positive incentives that encourage them to pay their fees. Many possibilities exist. Specific financial incentives might be developed, directed either at individual users (such as discounts for prompt payment) or at groups of users (such as allowing a water users' organisation to retain, for its own use, a portion of fees that it collects from its members). Incentives to pay may also be created through social factors. These may also be directed at the level of either the individual water user or a group of users. An example of the former would be having collections in a village handled through direct contacts between the individual farmers and a well-known and respected local person. Under such a collection arrangement, farmers may feel an increased obligation to pay their fees. An example of the latter is a system of awards for villages or water users' organisations that achieve target levels of collections. Incentives may also be established through institutional arrangements, such as earmarking fees collected in a particular project for use in that project, or establishing a mechanism whereby water users have a voice in expenditure decisions.

Both equity in the assessment of fees and positive incentives to encourage payment of fees are important; however, they need to be backed up by an effective means of enforcement. Although termination of water deliveries is potentially the most effective penalty for non-payment, in many situations such an approach is not feasible. Financial penalties may be imposed for late payment of fees; however, this approach can be effective only if the penalty can be enforced. In some cases social sanctions may be a very effective means of enforcement. Ultimately, it may be necessary to resort to legal means of enforcement. But if implementation of legal penalties is too cumbersome, the threat of their use may carry little credibility. This problem has sometimes been overcome through the development of simplified legal procedures for enforcing irrigation fees.

Finally, we note that with respect to all three of the above considerations affecting the ability to collect fees (equity in assessment, positive incentives to encourage payment, and effective enforcement mechanisms) decentralisation through devolution of the operational responsibilities for irrigation from the national level toward the individual project level can offer significant advantages. Such decentralised irrigation agencies are in a better position to learn of, and to respond to, local factors that affect the water users' perceptions of equity. By contrast, highly centralised agencies, which often must establish a structure of fees that will apply to irrigation projects over an entire region of a nation, or even over an entire nation, may be unable to respond to the different equity concerns of users in individual projects. Financial incentives to encourage payment of fees can be implemented by either centralised or decentralised irrigation agencies; however, for other types of incentives, the specific opportunities and details are likely to vary from one location to another. Decentralised agencies are in a much better position to identify these opportunities and to take advantage of them in developing specific incentives for payment. Furthermore, the greater the degree of decentralisation, the greater the assurance the users will have that the fees they pay will be spent to benefit the project serving them. The benefits of decentralisation with respect to the development of effective penalties for enforcement are similar to the case of incentives for payment. With the exception of financial penalties, local knowledge is likely to facilitate the establishment of an effective system of penalties for failure to pay the irrigation fees.

We end with a reminder of the caution that we offered early in the book. Deciding on the best approach to irrigation financing in any given context requires careful consideration of many factors specific to that context. Rational decisions about changes in financing policies cannot be made without reviewing the broader context of the nation's sectoral price and taxation policies, its general macroeconomic policies, its institutional context, its political environment, and its past experience with financing policies. No simple and universal answer can be given regarding the best approach to financing the recurrent costs of irrigation. But the framework of analysis that we have developed in this book should provide a strong foundation for examining financing policies in any given situation, and for making recommendations for their improvement.

Notes

Chapter 1

1 Repetto, R. (1986). *Skimming the Water: Rent Seeking and the Performance of Public Irrigation Systems*. Washington, D.C.: World Resources Institute, p. 4.
2 Cited by Whitcombe, E. (1972). *Agrarian Conditions in North India. The UP Under British Rule, 1860–1900*. Berkeley: University of California Press.
3 Starr, J. R. and D. C. Stoll (1987). *U.S. Foreign Policy on Water Resources in the Middle East*. Washington, D.C.: The Center for Strategic and International Studies.

Chapter 2

1 Krutilla, J. V. and O. Eckstein (1958). *Multiple Purpose River Development: Studies in Applied Economic Analysis*. Baltimore: The Johns Hopkins University Press.
2 This use of the term 'efficiency' is consistent with its broad dictionary definition of producing desired effects without waste. The concept of economic efficiency thus implies a *simultaneous* assessment of both the effects and the costs of a set of activities. In this sense, the ability to be effective in producing desired results is implied in the concept of efficiency. In some fields, however, such as those dealing with public administration and organisational behaviour, the term 'efficiency' is frequently used more narrowly to focus on cost, or the lack of waste. In this body of literature, the term 'effectiveness' is used to refer to the achievement of desired results, particularly in terms of broad social impacts; however, it does not incorporate any consideration of costs incurred.
3 Musgrave, R. A. (1959). *The Theory of Public Finance*. New York: McGraw-Hill.
4 This discussion is adapted from Bird, R. M. (1976). *Charging for Public Services: A New Look at an Old Idea*. Canadian Tax Papers No. 59. Toronto: Canadian Tax Foundation.

Chapter 3

1 As noted in Chapter 2, we are using the term 'efficiency' in its broad meaning of producing desired effects without waste.
2 Bowen, R. L. and R. A. Young (1986). 'Appraising alternatives for allocating and cost recovery for irrigation water in Egypt', *Agricultural Economics: The Journal of the International Association of Agricultural Economists*, **1**, 35–52.
3. The data from which this example is derived are from Chaudhry, M. A. (1987). 'Irrigation water charges and recurrent cost recovery in Pakistan', in *Technical Papers from the Expert Consultation on Irrigation Water Charges: Volume II*. Rome: Food and Agriculture Organization of the United Nations and the United States Agency for International Development, pp. 19–36.

Chapter 5

1 One danger of spreading water to increase area in arid zones is that it may increase the long-term process of soil salinisation. A tenant farmer with no long-term interest in the land, or a low-income farmer with a high preference for current over future income, may find these long-term financial costs to be acceptable. However, the economic costs to society of this erosion of productive potential will be much higher.
2 Confusion is sometimes created when the term 'marginal cost pricing' is used in ways that essentially mean the same as is meant by the term 'volumetric pricing' or simply 'pricing'. Anyone using the term in this fashion is focusing on the fact that true pricing (whether achieved by volumetric measurement or by some proxy for volume) gives water a positive marginal cost *to the water user*. This is, of course, true regardless of the level at which the price is set. But the term 'marginal cost pricing' more properly refers to a particular level of price – namely a price set at the marginal cost *to society* of supplying water to the users.

Chapter 6

1 Moris, J. (1987). 'Irrigation as a privileged solution in African development', *Development Policy Review*, **5**, 99–123.
2 Joss, A. (1945). Repayment experience on federal reclamation projects', *Journal of Farm Economics*, **27**, 153–167.
3 United States Water Resources Policy Commission (1950). 'A water policy for the American people'. Washington, D.C.: Report of the President's Water Resources Policy Commission, p. 76.
4 Wade, R. (1982). *Irrigation and Agricultural Politics in South Korea*. Boulder, Colorado: Westview Press.

Chapter 7

1 Wolf, J. M. (1985?). 'Cost and financing of irrigation system operations and maintenance in Pakistan' (typescript), and Development Alternatives, Inc. (1985). *Funding Requirements for Adequate Irrigation System Operation and Maintenance – Pakistan*. Report for the U.S. Agency for International Development Mission to Pakistan by W. L. McAnlis, W. H. Rusk and J. M. Wolf.

2 Bhatia, R. (1989). 'Financing irrigation services in India: a case study of Bihar and Haryana States, in L. E. Small *et al.*, *Financing Irrigation Services: A Literature Review and Selected Case Studies from Asia*. Colombo, Sri Lanka: International Irrigation Management Institute, p. 274.

3 Browning, E. (1976). 'The marginal cost of public funds', *Journal of Political Economy*, **84(2)**, 283–98; Stuart, C. (1984). 'Welfare costs per dollar of additional tax revenues in the United States', *American Economic Review*, **74(3)**, 352–62.

Chapter 8

1 U.S. Department of Interior, Bureau of Reclamation (1981). *Draft Environmental Impact Statement on Acreage Limitation: Westwide Report*. Washington, D.C., cited in Repetto, R. (1986). *Skimming the Water: Rent-seeking and the Performance of Public Irrigation Systems*. World Resources Institute, Research Report No. 4, p. 18.

2 It is true that for a specific irrigation project, many of the indirect benefits (particularly those linked to the marketing of inputs and outputs) tend to be concentrated in the geographic region of the project. But a government's irrigation investment programme would typically involve a considerable number of individual projects widely distributed throughout the nation. Thus the distribution of the indirect benefits of irrigation is likely to be much broader than the distribution of the benefits of any one specific irrigation project.

Chapter 10

1 Svendsen, M. and E. Lopez (1980). 'The Talaksan pump irrigation project'. *The Determinants of Developing-Country Irrigation Project Problems*, Contract no. AID/ta-C-1412 between the U.S. Agency for International Development and Cornell University, Technical Report No. 1. Ithaca, N.Y.: Cornell University.

Chapter 11

1 Abel, M. E. (1976). 'Irrigation systems in Taiwan: decentralized irrigation management', *Water Resources Research*, **12(3)**, 346.

2 *Ibid.*, pp. 346–7.

3 Bagadion, B. U. (1988). 'The evolution of the policy context: an historical overview', in F. F. Korten and R. Y. Siy, Jr (eds), *Transforming a Bureaucracy: The Experience of the Philippine National Irrigation Administration*. West Hartford, Connecticut: Kumarian Press, Inc., p. 8.

4 Maass, A. and R. L. Anderson (1978). *. . . and the Desert Shall Rejoice: Conflict, Growth and Justice in Arid Environments*. Cambridge, Mass.: MIT Press.

5 Wade, R. (1982). *Irrigation and Agricultural Politics in South Korea*. Boulder, Colorado: Westview Press.

Index